KU-472-963

Scientific Basis of Healthcare

Asthma

Scientific Basis of Healthcare

- AIDS and Pregnancy
- Angina
- Arthritis
- Asthma

Editorial Advisory Board

Prof. Sally Wai-chi Chan (Singapore)
Prof. Caroline J. Hollins-Martin (UK)
Prof. Diana T.F. Lee (Hong Kong)
Dr. Jean Rankin (UK)

Scientific Basis of Healthcare
Asthma

Editors

Colin R. Martin PhD
Chair in Mental Health, School of Health
Nursing and Midwifery
University of West of Scotland
UK

Victor R. Preedy PhD DSc
Professor of Nutritional Biochemistry
School of Medicine
King's College London
and
Professor of Clinical Biochemistry
King's College Hospital
UK

Science Publishers
Jersey, British Isles
Enfield, New Hampshire

Published by Science Publishers, an imprint of Edenbridge Ltd.

- St. Helier, Jersey, British Channel Islands
- P.O. Box 699, Enfield, NH 03748, USA

E-mail: info@scipub.net Website: www.scipub.net

Marketed and distributed by:

6000 Broken Sound Parkway, NW
Suite 300, Boca Raton, FL 33487
270 Madison Avenue
New York, NY 10016
2 Park Square, Milton Park
Abingdon, Oxon OX14 4RN, UK

Copyright reserved © 2012

ISBN 978-1-57808-731-0

Cover Illustrations: Reproduced by kind courtesy of the undermentioned authors:
- Figure No. 1 from Chapter 6 by Hannah K. Bayes and Neil C. Thomson
- Figure No. 2 from Chapter 9 by Pippa Hall and Andrew Bush

Library of Congress Cataloging-in-Publication Data

Scientific basis of healthcare asthma / editors, Colin R. Martin, Victor R. Preedy.
p. cm.
Includes bibliographical references and index.
ISBN 978-1-57808-731-0 (hardback)
1. Asthma. 2. Nurses. 3. Allied health personnel. I. Martin, Colin R., 1964- II. Preedy, Victor R.
RC591.S333 2012
616.2'38--dc23

2011038879

The views expressed in this book are those of the author(s) and the publisher does not assume responsibility for the authenticity of the findings/conclusions drawn by the author(s). No responsibility is assumed by the publisher for any injury and/or damage to persons or property as a matter of products liability, negligence or otherwise, or from any use or operation of any methods, products, instructions or ideas contained in the material herein. Because of rapid advances in the medical sciences, in particular, independent verification of diagnoses and drug dosages should be made.

All rights reserved. No part of this publication may be reproduced, stored in a retrieval system, or transmitted in any form or by any means, electronic, mechanical, photocopying or otherwise, without the prior permission of the publisher, in writing. The exception to this is when a reasonable part of the text is quoted for purpose of book review, abstracting etc.

This book is sold subject to the condition that it shall not, by way of trade or otherwise be lent, re-sold, hired out, or otherwise circulated without the publisher's prior consent in any form of binding or cover other than that in which it is published and without a similar condition including this condition being imposed on the subsequent purchaser.

Printed in the United States of America

Foreword

Asthma is a long-term condition involving airflow obstruction due to a narrowing of the airways. Increasing knowledge in recent decades about the pathology of asthma has made it apparent that environmental and behavioural management are both equally important as medical management. Surprisingly common though this condition is, asthma-related deaths are not only exceptional but, in many cases, preventable. Furthermore, prevention of acute episodes involves a large degree of behavioural change. Consequently, the emotional burden of living with asthma can impact greatly on the individual's quality of life, indeed equally as much as the physical limitations and complications associated with this condition, as well as impacting significantly on their family and friends. There is good evidence that the experience of asthma may precipitate both anxiety and depression, thus a comprehensive and holistic understanding of this enduring presentation is essential to provide contextually-specific understanding, care and effective clinical management.

The complex interacting factors implicated in the genesis and course of asthma require understanding that is primarily achieved by undertaking high quality research, the results of which can then be translated into evidence-based practice. My own research in the area of chronic disease management has highlighted to me that many of these issues are common across different conditions, whether they be physical limitations such as sleep difficulties, pain, difficulty breathing or psychological impact, such as stress, anxiety and depression. Thus, chronic disease provides a context that facilitates dynamic interaction between these seemingly unrelated domains—the physical and the psychological—the result of which is the unique and personal experience of pathology by the individual. Recognising that research continually broadens our horizons, offering many innovative and effective opportunities to improve asthma management in the future, a pressing matter for all health professionals is how to remain up to date with comprehensive and relevant information. Colin Martin and Victor Preedy are therefore to be applauded in bringing together an impressive group of experts in asthma to produce this comprehensive evidence-based volume. There is no doubt that this will be welcomed by those working in a research environment, clinical practice or indeed both. It is a pleasure, therefore, to recommend this text for those working in this area.

Professor Jane Speight

Foundation Director, The Australian Centre for Behavioural Research in Diabetes, a partnership for better health between Diabetes Australia—Victoria and Deakin University Melbourne, Australia

Professor Jane Speight is a health psychologist and researcher who has worked extensively in the area of chronic disease and its management. She is the Foundation Director of The Australian Centre for Behavioural Research in Diabetes, a partnership for better health between Diabetes Australia—Victoria and Deakin University, Melbourne, Australia, where she holds the Chair in Behavioural and Social Research in Diabetes. Professor Speight is actively involved internationally in many clinical research studies aiming to improve the lives of people affected by chronic disease.

Preface

In the past three decades there has been a major sea change in the way healthcare is taught and implemented. Teaching in the healthcare professions have been replaced from a "this is what you do" approach to "this is the scientific basis" ethos of evidence-based material. The healthcare professional is now more educated, more informed and more aware that the foundation of good health is good science. Healthcare practitioners with research doctorates or Masters degrees are also more commonplace. As a consequence, practice and procedures are continually changing with a corresponding improvement in healthcare. Concomitantly, the demand for comprehensive and focused evidenced-based text and scientific literature covering single areas of healthcare or treatments have also increased. Hitherto, these have been difficult to obtain and thus it was decided to work on a collection of books on **The Scientific Basis of Healthcare**. The chapters impart holistic information on the scientific basis of health and covers the latest knowledge, trends and treatments. The ability to transcend the intellectual divide is aided by the fact that each chapter has:

- An *Abstract*
- A section called *"Practice and Procedures"* where relevant
- *Key Facts* (areas of focus explained for the lay person)
- *Definitions of words and terms*
- *Summary points*

The books, each on a different medical condition, cover a wide number of areas. The chapters are written by national or international experts and specialists.

In **Asthma** we cover a wide range of areas and subtopics including overviews and introductory text, quality of care, prostaglandins and other mediators of inflammation, unsatisfactory asthma control, rthinosinusitis, psychological distress, adherence to guidelines.

inhalation techniques in asthma, nurse-led home visits, therapy in elderly patients, pediatric asthma and many more scientific fields associated with **Asthma**.

The books are designed for practicing health care workers, trained nurses, nursing students, doctors and medical students, therapists, trainees and practitioners of all health-related disciplines including physiotherapists, midwives, dietitians, psychologists, and so on. The special feature of the book also means it is suitable for post graduates, special project students, teachers, lecturers and professors. It is also suitable for college, university, nursing, and medical school libraries as a reference guide.

Colin R. Martin
Victor R. Preedy

Contents

1

The Quality of Asthma Care within Primary Care

Mark Karaczun[1,a] Asmaa Abdelhamid,[1,b,*] Sue Maisey[1,c] and Nicholas Steel[1,d]

ABSTRACT

Primary care (PC) is an important source of healthcare for people with asthma, a chronic respiratory disease causing intermittent airway inflammation. Many models have been applied to measure the quality of PC including auditing routine practice against evidence-based clinical guidelines. Recent literature shows that the degree to which practices achieve clinical guideline criteria varies widely, often falling short of ideal standards. Reportedly, less than one-half of patients are assessed with spirometry. Inhaled corticosteroids (ICS) are often not prescribed when they should or are prescribed at an inappropriate dose. Assessment of inhaler technique is far from universal (14–68%). Written personal asthma action plans are only given in about one-quarter of cases (12–44%). Reasons identified for this include the need for flexibility in treatment. Incentive schemes may improve performance against guidelines. Factors impeding clinician adherence to guidelines

[1]School of Medicine, Health Policy and Practice, Faculty of Health, University of East Anglia, Norwich, Norfolk NR4 7TJ, United Kingdom.

[a]Email: karaczun@doctors.org.uk

[b]Email: asmaa.abdelhamid@uea.ac.uk

[c]Email: s.maisey@uea.ac.uk

[d]Email: n.steel@uea.ac.uk

*Corresponding author

List of abbreviations after the text.

included complicated or multiple versions of guidelines, limited skills or consultation time to carry out patient education, and disagreement with the usefulness of specific guideline criteria. Inferior quality of care and/or poor asthma outcomes has been associated with factors such as female gender, older age, and rurality. Shortfalls against guideline criteria may be due to fundamental differences between the demands of PC and the artificial research conditions upon which guidelines are based.

INTRODUCTION

Most healthcare for asthma patients takes place within primary care (PC) (Moth et al. 2008). Unlike specialists who can focus on one particular disease, PC clinicians must optimize the management of a person's asthma despite competing demands from the patient's co-morbidities and complex psycho-social factors. This chapter examines the quality of such care. Section One defines asthma and PC, and briefly outlines common approaches to quality measurement. Section Two presents key findings from recent published literature on the quality of care for asthma patients within PC practice and discusses potential reasons for sub-standard performance against clinical guidelines.

SECTION ONE: DEFINITIONS AND MODELS

Asthma is a chronic inflammatory disorder of the airways affecting 300 million people (GINA 2009). It is characterized by episodic constriction of airway smooth muscles resulting in largely reversible airflow obstruction. Clinical diagnosis is far from straightforward as typical symptoms of wheeze, cough, and chest tightness are non-specific.

Primary Care (PC) systems vary widely between countries and are sparse in some areas. For the purposes of this chapter, PC includes 'General Practice' (GP) as in the United Kingdom, 'Family Medicine' as in the United States, and other models like community-based pediatrics. The term 'GP' refers to any PC doctor. 'Clinician' also includes nurses.

Elements generally considered fundamental to PC (Kringos et al. 2010, Starfield 2009) include:

- *First contact*: Patients generally first seek clinical advice, assessment, and/or treatment for new problems via PC.
- *Longitudinality*: Unlike specialists who may discharge a patient from their care once a specific problem has resolved, GPs maintain relationships with their patients over time.
- *Person-centeredness/Comprehensiveness*: GPs are consulted for a broad range of physical and mental health issues.
- *Coordination*: GPs are frequently gatekeepers to specialist services.

Quality of healthcare can vary in place and time. Moreover, a patient, their GP, and a local health economist might have different perspectives on what constitutes high quality care. Quality frameworks such as those from the Institute of Medicine (IOM 2001) and the Office for Economic Cooperation and Development (OECD) (Kelley and Hurst 2006) have been developed from Donabedian's triad of structure, process and outcome and Maxwell's dimensions of effectiveness, efficiency, acceptability, access, equity, and relevance (Donabedian 1980, Maxwell 1992).

Healthcare quality can be improved in three broad ways (Steel et al. 2007a):

- Regulation and compulsory standards: e.g., clinician licensing.
- Education and clinical audit: e.g., knowledge of current best practice, self-testing of the extent of its application by systematic methods and action to improve where practice falls short (Burgess 2011).
- Market and financial factors: e.g., the publication of league tables to enable patient choice of provider.

Quality improvement within PC depends on the availability of robust evidence-based indicators of care, ideally developed using systematic techniques and embodying face/content validity, feasibility and reproducibility of measurements, sensitivity to change, ability to predict good outcomes and, crucially, acceptability to those whose practice is being assessed (Campbell et al. 2004). While indicators may measure structure, process or outcomes of care, in practice, reliable outcome measurement can be very difficult due to confounding factors unrelated to healthcare. Indicators of processes of care, which describe actual practice, are thought to be more likely to lead to quality improvement (Kelley and Hurst 2006, Campbell et al. 2004). Such indicators can form the basis of asthma clinical guidelines (To et al. 2010) which are usually published nationally, but can also be developed locally or internationally (Table 1).

Measuring patient-centeredness is complicated by various elements such as consultation skills and shared decision-making classified under this broad term (Howie et al. 2004). A systematic review (Sans-Corrales et al. 2006) of general aspects of patient satisfaction indentified various factors important to patients including:

- Being treated with respect and dignity.
- Obtaining an appointment quickly (same or following day).
- Short waiting times (<10 min).
- Longer consultations (10 min versus 5 min).
- Friendly confident clinicians who are attentive, provide clear explanations and share treatment decisions.
- Continuity of care with the same clinician over time.

Such factors are commonly gauged via patient satisfaction questionnaires.

In summary, the quality of asthma care within PC can be audited against agreed clinical guideline criteria or gauged by surveys.

SECTION TWO: ASTHMA CARE QUALITY WITHIN PRIMARY CARE

This section summarizes key findings from studies on PC-based quality of asthma healthcare published since 2000. Studies exclusive to secondary care or clinical trials were excluded. Findings are presented under the IOM domains of Effectiveness, Safety, Patient-centeredness, and Equity. The former are grouped under themes paralleling a patient's journey through PC.

Standards for high quality asthma care, broadly represented by the following, naturally evolve over time and vary according to local guidelines. Diagnosis should be accurate, timely, and documented. Patients should attend regular structured review clinics during which the severity and control of the patient's asthma should be assessed on symptoms and lung function measurements. The appropriate level of treatment should be prescribed alongside assessment of

Table 1 Examples of Asthma Clinical Guidelines Used within Primary Care.

Country	Year	Organisation(s)	Additional Reference Information
International	2009	GINA: Global Initiative for Asthma	Global Strategy for Asthma Management and Prevention. Medical Communications Resources, Inc. www.ginasthma.org
International	2008	International Union Against Tuberculosis and Lung Disease ("The Union")	Aït-Khaled, N. and D.A. Enarson, C.Y. Chiang, G. Marks and K. Bissell. Management of asthma: a guide to the essentials of good clinical practice, 3rd Ed. Paris, France. www.theunion.org
Australia	2006	National Asthma Council	Asthma Management Handbook. South Melbourne, Australia www.nationalasthma.org.au
Canada	2010	Canadian Thoracic Society	Lougheed M.D., C. Lemiere, S.D. Dell, F.M. Ducharme, J.M. Fitzgerald, R. Leigh, C. Licskai, B.H. Rowe, D. Bowie, A. Becker and L.P. Boulet. Asthma Management Continuum—2010 Consensus Summary for children six years of age and over, and adults. Can Respir J 17: 15–24; www.respiratoryguidelines.ca
India	2005	World Health Organization (WHO)—Government of India Collaborative Programme.	Jindal S.K. and D. Gupta, A.N. Agarwal and R. Agarwal. Guidelines for Management of Asthma at Primary and Secondary Levels of Health Care in India. Chandigarh, India. www.indiachest.org/asthama_guidelines.html
Ireland	2008	Irish College of General Practitioners (ICGP)	Holohan, J. and P. Manning. 2008. Asthma Control in General Practice. ICGP Quality in Practice Committee, Dublin, Ireland
United Kingdom	2009	British Thoracic Society (BTS) and Scottish Intercollegiate Guidelines Network (SIGN)	British Guideline on the Management of Asthma: A National Clinical Guideline. ISBN: 978 1 905813 28 5 www.brit-**thoracic**.org.uk www.sign.ac.uk/guidelines/ fulltext/101/index.html
USA	2007	National Heart, Lung and Blood Institute (NHLBI) and National Asthma Education and Prevention Program (NAEPP)	Expert Panel Report 3: Guidelines for the Diagnosis and Management of Asthma. National Institutes of Health ,U.S. Department of Health and Human Services, Bethesda, USA. NIH Publication No. 07-4051 http://www.nhlbi.nih.gov/guidelines/asthma/

The above are examples of commonly used asthma guidelines. There may also be other guidelines local to your area or for different age groups.

the patient's inhaler technique, understanding of asthma, and exacerbating factors. This should form the basis of a tailored written Personal Asthma Action Plan (PAAP) which allows them to self-adjust their treatment.

Effectiveness and Safety

Identifying and tracking asthma patients

Patients with a confirmed asthma diagnosis should be included in a database to facilitate identification and sampling (Nizami and Mash 2005). Though >90% of practices in Scotland had an asthma register, only 76% updated these in the previous three years and 63% used them to conduct automated system searches for at-risk patients (Hoskins et al. 2005). Only 4% of local Swedish practice patients had a diagnosis of asthma despite national prevalence rates of 8% (Weidinger et al. 2009), suggesting under-diagnosis.

Proxy indicators such as documentation of wheeze or inhaler prescriptions can help identify asthma cases. One Swedish study showed that nearly half (46.5%) of patients receiving anti-asthma medication lacked a recorded asthma or alternative diagnosis (Weidinger et al. 2009). Others reported that having a recorded diagnosis was higher for specific groups of patients such as children, older adults, and those with increasing peak expiratory flow (PEF) rates (Aït-Khaled et al. 2006).

Symptoms and initial assessment

Accurate assessment of patient symptoms is critical to asthma diagnosis and monitoring which may be uncontrolled in nearly 75% of cases (Carlton 2005). Key information such as wheeze or smoking status has been recorded in as few as 10% of cases (Bateman et al. 2009). Doctors often underestimate the severity of asthma or overestimate its control particularly when relying on personal clinical judgment rather than assessing symptoms against guideline criteria (Aït-Khaled et al. 2008, Chapman et al. 2008).

One American study found that >50% of asthmatics presenting to their GP for any reason had uncontrolled asthma at the time, including nearly half of those presenting for non-respiratory complaints (Mintz et al. 2009). This suggests opportunistic assessment of asthma during any PC visit may be beneficial. Recording of criteria such as risk factors and symptoms has been enhanced by tailored medical record systems (Abudahish and Bella 2010) and audit with benchmarking (Silver et al. 2011).

Investigations

While spirometry is superior for diagnosing asthma, in practice many GPs rely on PEF measurements to monitor lung function as it is easier and more readily available (Neville et al. 2004, Hasselgren et al. 2005). Low rates (e.g., 44%) of spirometry may be attributable to GP perception that it is useful only in patients with more troublesome symptoms (Chapman et al. 2008). Adults reported lung function testing more than children albeit the latter may not be able to perform the test (Weidinger et al. 2009).

Pharmacological treatment

Asthma treatment is based on a step-wise protocol (Fig. 1) wherein the types of medication and dosing change according to current levels of symptom control. However, GPs may not always prescribe according to asthma guidelines. For example, a study of German practices found such accordance in only 37% of cases particularly with respect to inhaled corticosteroids (ICS) (Schneider et al. 2007). Though this may be due to the lack of availability of ICS inhalers in some areas, even where plentiful, they may be under-prescribed (Aït-Khaled et al. 2006, Abudahish et al. 2010). Similarly, only 24% of patients in rural Scotland had a supply of oral corticosteroid tablets on hand to take in case of an emergency (Wiener-Ogilvie et al. 2007).

It is not sufficient to prescribe the correct medicine. It must also be prescribed at an appropriate dose. One international study found over 10% of those patients receiving ICS were given prescriptions more than one step below what they required and 5% two steps too high. Under-dosing was associated more with using PEF rather than symptoms to judge the severity of asthma (Aït-Khaled et al.

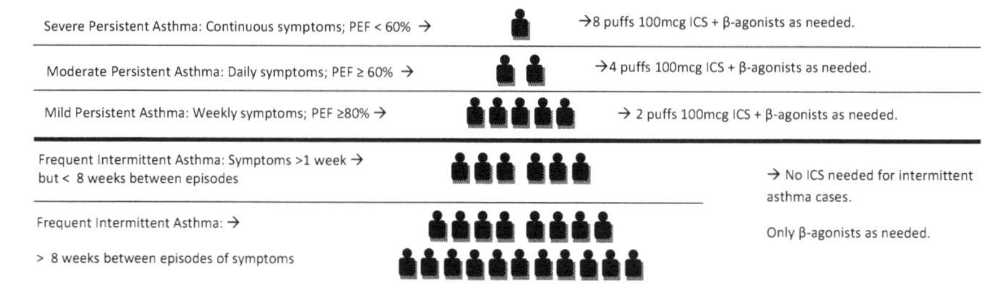

Figure 1 Stepwise Treatment Regimen Example.
Patient symptoms and/or Peak Expiratory Flow (PEF) readings as outlined in the left hand column determine treatment as per the example in the right hand column. The middle column shows the approximate proportion of asthma patents typically found at each step. This is only an example and should not be used as a basis for treatment in lieu of your local guidelines. Adapted from the 2008 Union Guidelines and 2006 Australian Guidelines as listed in Table 1. = one case; ICS= Inhaled corticosteroids.

2006). Qualitative studies have identified a common preference amongst GPs to err on the side of over-treating rather than under-treating as control of acute asthma symptoms was perceived as being more important than avoiding questionable long-term risk of side-effects (Goeman et al. 2005). Receipt of asthma medication does not necessarily equate to proper asthma care as sometimes prescriptions can be refilled without seeing a GP. A Danish study found 34% of children had no asthma-related contacts with health services aside from receiving prescriptions for asthma medications (Moth et al. 2008).

Inhaler technique

Other than systemic medication in severe cases, most PC asthma drugs are inhaled. Commonly, this is via pressurized metered dose inhalers (pMDIs) in which actuation of the pMDI must be properly coordinated with inhalation. It is important to assess patients' inhaler technique, discussed in more detail in a later chapter, since improper technique reduces the amount of medication that is delivered to the airways where they exert their effect. Reports of how routinely this is assessed vary widely (14–68%) but have been shown to improve (initially 55% then 67%) after audit (Bateman et al. 2009, Neville et al. 2004, Nizami and Mash 2005, Chapman et al. 2008).

Adherence to prescribed treatment

Reasons for poor patient adherence to treatment are complex but include avoiding taking medication due to concerns about long-term side-effects or believing that the medication is unnecessary (Barnes 2004). Greater adherence has been associated with patient understanding that ICS is a preventive therapy (Nizami and Mash 2005); by shared decision-making between doctors and patients (Arbuthnott and Sharpe 2009); and higher cultural competence by, and patient satisfaction with, PC practices (Lieu et al. 2004).

Education and review

Asthma guidelines increasingly stress the importance of patient education. When coupled with regular medical review, educating patients in self-management techniques which allowed them to adjust their medication according to a pre-determined plan (e.g., PAAP) improves outcomes (Gibson et al. 2009). Nearly 75% of asthma patients in eastern Germany reported failing to receive any asthma education. However, 64% indicated they did not want education (Schneider et al. 2007).

Guidelines recommend regular structured review clinics for assessment of symptom control, inhaler technique, and lung function so management can be adjusted accordingly. The use of trained asthma nurses has become increasingly common in order to assist with some of these tasks or, increasingly, to lead on the management of asthma patients (Hoskins et al. 2005)—a development often welcomed by GPs (Goeman et al. 2005). Others have reported success with different models of review such as holding clinics external to the practice (Lyte et al. 2007).

Written Personal Asthma Action Plans (PAAPs)

PAAPs (Fig. 2) are recommended in asthma guidelines particularly for patients with moderate to severe asthma as one hospital visit can be avoided for every eight PAAPs administered (Gibson et al. 2009, Powell & Gibson 2009). In principle, these allow patients to adjust their medication according to pre-determined criteria.

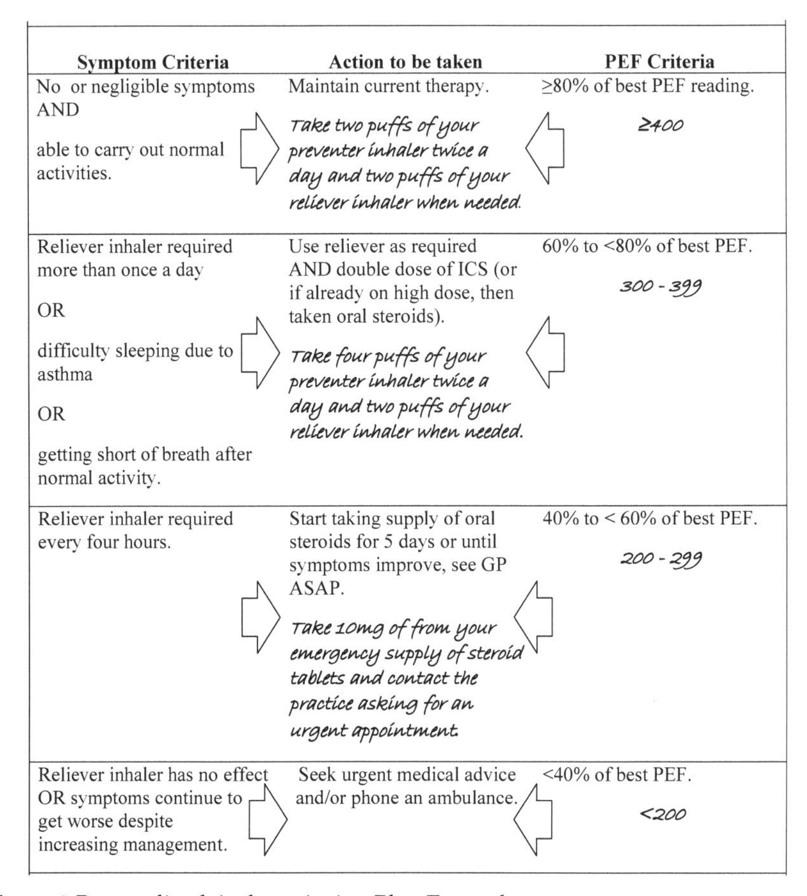

Symptom Criteria	Action to be taken	PEF Criteria
No or negligible symptoms AND able to carry out normal activities.	Maintain current therapy. *Take two puffs of your preventer inhaler twice a day and two puffs of your reliever inhaler when needed.*	≥80% of best PEF reading. *≥400*
Reliever inhaler required more than once a day OR difficulty sleeping due to asthma OR getting short of breath after normal activity.	Use reliever as required AND double dose of ICS (or if already on high dose, then taken oral steroids). *Take four puffs of your preventer inhaler twice a day and two puffs of your reliever inhaler when needed.*	60% to <80% of best PEF. *300 - 399*
Reliever inhaler required every four hours.	Start taking supply of oral steroids for 5 days or until symptoms improve, see GP ASAP. *Take 10mg of from your emergency supply of steroid tablets and contact the practice asking for an urgent appointment.*	40% to < 60% of best PEF. *200 - 299*
Reliever inhaler has no effect OR symptoms continue to get worse despite increasing management.	Seek urgent medical advice and/or phone an ambulance.	<40% of best PEF. *<200*

Figure 2 Personalized Asthma Action Plan Example.
This is only an example and not intended for actual clinical use. Actual plans will differ according to guidelines and the patient's needs. For example, some plans may have only two or three steps. Some guidelines currently stress basing treatment on symptoms rather than peak flow readings, particularly in children. However, plans based on PEF are still common. Normal text basic refers to generic principles. Stylized text represents an example of personalized criteria for an adult with best PEF reading of 500L/min.

Despite this, a minority of asthma patients receive PAAPs, with rates being remarkably similar in studies from different countries (Table 2). Despite 80% of Scottish GPs thinking that implementation of PAAPs would lead to better outcomes, only 13% admitted normally providing them and only 23% of their patients reported receiving such a plan (Wiener-Ogilvie et al. 2007).

Table 2 Percentage of Asthma Patients Having a Written Personal Asthma Action Plan.

%	Location	Reference	Comments
12%–28%	USA	Silver et al. 2010	12% pre-quality improvement initiative; 28% post-initiative.
20%	Canada	Chapman et al. 2008	Survey of over 10,000 asthma patients attending their GP for any reason.
23%	France	Laforest et al. 2006.	16% for those solely under GP care; 30% solely under specialist care.
23%	Scotland	Wiener-Ogilvie et al. 2007	Survey of patients; 58 of 254 respondents.
23%–44%	Australia	Glasgow et al. 2003	Intervention group 23% at baseline improved to 44% after twelve months; Control group 28% to 34% respectively.
43%	USA	Reeves et al. 2006	No significant difference between urban, suburban, and rural children in Michigan.

These studies were striking for their consistency across countries showing that only about a quarter of asthma patients received written Personal Asthma Action Plans (PAAPs). These may also be referred to as Written Action Plans.

Supplementary References:

Glasgow, N.J., A.L. Ponsonby, R. Yates, J. Beilby and P. Dugdale. 2003. Proactive asthma care in childhood: general practice based randomised controlled trial. BMJ 327: 659–665.

Laforest, L., E. Van Ganse, G. Devouassoux, S. Chretin, L. Osman, G. Baugil, Y. Pacheco and G. Chamba. 2006. Management of asthma in patients supervised by primary care physicians or by specialists. Eur Respir J 27: 42–50.

Reeves, M.J., S.R. Bohm, S.J. Korzeniewski and M.D. Brown. 2006. Asthma Care and Management Before an Emergency Department visit in children in Western Michigan: How Well does Care Adhere to Guidelines? Pediatrics 117: S118–S126.

Ring et al. (2007) reviewed various methods used to promote PAAPs. Although these included mostly hospital-based interventions, some methods such as telephone calls from an asthma educator or internet-based monitoring have potential for PC. Proactive measures such as prompts to attend asthma review improved uptake. Making repeat prescriptions dependent on attending regular review and receipt of PAAPs improved PAAP rates from 30% to 65% in Australia (Byrnes et al. 2010).

Clinician perceptions of PAAPs

Qualitative studies (Wiener-Ogilvie et al. 2007, Goeman et al. 2005, Douglass et al. 2002) have identified clinician views which may make PAAPs unpopular:

- Uncertainty about their content or usefulness and perceived inflexibility of generic PAAPs as inappropriate for a dynamic condition like asthma.
- GPs feeling de-skilled by asthma nurses. Nurses feeling unsupported by GPs.
- Fear that patients might adhere to outdated PAAPs, find them confusing, or feel discouraged from attending regular asthma review.
- Medico-legal concerns regarding committing specific advice to writing.

Patient perceptions of PAAPs

Patients may lack confidence in using PAAPs, some relying, instead, on informal strategies for managing exacerbations (Douglass et al. 2002). Many felt PAAPS were complicated, bothersome, or not relevant to them. Pre-printed versions with images of children on the cover likely discouraged use by adults (Silver et al. 2011).

Barriers and Enabling factors to guideline adherence

Adherence to guidelines can be limited by many factors (Table 3). In addition to the examples already mentioned, asthma guideline adherence can also be improved by use of decision tools at the point of care (Carlton et al. 2005) and linking standards (Table 4) to financial incentives (Steel et al. 2007b). Though asthma care quality was already improving beforehand, the rate of improvement increased after the introduction of pay-for-performance incentives (Campbell et al. 2007).

Table 3 General Barriers Impeding Adherence to Asthma Guidelines.

Factor	Asthma-related example(s)	References
Lack of awareness of or familiarity with guidelines.	GPs may use outdated versions or be faced with multiple guidelines for asthma and other diseases competing for their attention.	Abudahish and Bella. 2007, Lyte et al. 2007
Lack of agreement with the content or relevance of specific guideline recommendations.	Belief that spirometry is not cost-effective in asymptomatic patients with asthma.	Silver et al. 2011
Lack of self-efficacy in applying them.	GPs may feel de-skilled in conducting spirometry which is often performed by asthma nurses.	Wiener-Ogilvie et al. 2007
Lack of outcome expectancy, e.g., "What's the point?"	Clinicians may feel that parents of asthmatic children will not heed advice to stop smoking.	Cabana et al. 2001.
External barriers which are patient-related	Some patients may shun medication, preferring non-medicinal approaches such as breathing exercises.	Mansour et al. 2000
External barriers related to practice-setting	Limited time for detailed consultation or counseling on factors such as smoking cessation. Lack of continuity of care with same GP.	Cabana et al. 2001, Lyte et al. 2007
Guideline-related barriers	Treatment algorithms may be viewed as too complicated.	Abudahish and Bella 2007

Adapted from the model proposed by Cabana M.D., C.S. Rand, N.R. Powe, A.W. Wu, M.H. Wilson, P.A.C. Abboud and H./R. Rubin. 1999. Why Don't Physicians Follow Clinical Practice guidelines? A Framework for Improvement. JAMA 282: 1458–1465.

Supplementary References:

Cabana, M.D., C.S. Rand, O.J. Becher and H.R. Rubin. 2001. Reasons for Pediatrician Nonadherence to Asthma Guidelines. Arch Pediatr Adolesc Med 155: 1057–1062.

Mansour, M., B.P. Lanphear and T.G. DeWitt. 2000. Barriers to Asthma Care in Urban Children: Parent Perspectives. Pediatrics 106: 512–519.

Table 4 Example of How Quality Indicators are Used in an Incentive Scheme for Primary Care.

Topic	What is measured	Including those with (codes for)	Excluding those with (codes for)	Timeframe
Register	The practice can produce a register of patients with asthma.	Recorded asthma diagnosis (Read code H33)	Resolved asthma	Within last fiscal year
			No prescription of asthma medications	In the previous 12 months
Reversibility measures	The percentage of patients with measures of variability or reversibility.	Spirometry measurements PEF measurements	Under age 8; Exceptions e.g. inability to carry out test.	Since 01 April 2006
Smoking	Percentage of patients with asthma with a record of smoking status.	On register	Under age 14; Over age 19	Previous 15 months
Review	Percentage of patients with asthma who have had an asthma review.	On register	Exception codes	Previous 15 months

This summarises the asthma criteria under the United Kingdom's Quality Outcomes Framework (QOF) incentive scheme for PC practices in the National Health Service (NHS). Practices meeting the targets are rewarded financially. An example of an exception code would be patients who are terminally ill or with severe co-morbidities which would deem calling them for asthma review inappropriate to their circumstances.

Table based on National Health Service IC—QOF Business Rules team (on behalf of DH). 2010. Asthma Ruleset Version No 18.0.

Patient-Centeredness

Qualitative studies (Dixon-Woods et al. 2002, Lyte et al. 2007) reported that young children with asthma felt their asthma should be closely monitored by clinicians showing keen interest in their progress. Children perceived themselves able to judge clinician competence. However, patient satisfaction surveys typically rely on parents to respond on behalf of their children, thus denying an adequate voice to both of them.

Access and Equity

Differences in asthma outcome may be due to differences in access or equity (Table 5) or may be due to biological, psychological, social, or institutional factors which may be wholly unrelated to PC. In particular, poorer outcomes or services have been associated with female gender, older age, and rurality.

Summary and Discussion

In summary, recent literature reveals wide variations in PC-based asthma care. There is room for improvement in most guideline criteria, particularly the routine use of PAAPs. There may be several fundamental reasons why PC has not performed better against guideline criteria.

First, PC may differ considerably from artificial research conditions. Notwithstanding any flaws, biases and limitations (Jadad et al. 2000) inherent to the trials and systematic reviews on which guidelines are based, even rigorous evidence is derived from highly controlled conditions using motivated volunteers who have been screened for factors which could confound results. Unlike the researcher who can carefully engineer their study population, GPs are not at liberty to exclude patients who can present with a wide variety of biological, psychological, and sociological issues which confound asthma care. Therefore, the patients in the PC-based studies highlighted here may differ considerably from those in the clinical trials which informed the guidelines used in these studies.

Table 5 Examples of Variations in Equity Concerning Asthma.

Attribute	Summary of findings	Reference
Age	Adults <35 had better control than adults >65. (OR 1.34)	Chapman et al. 2008
Gender	Being female is associated with poorer asthma control. (OR 0.87)	Chapman et al. 2008
Ethnicity	South Asian children in the UK paradoxically had fewer symptoms (7.6% v. 10.6%) but twice the hospitalization rates for asthma compared to white children	Netuvelli, G., B. Hurwitz, M. Levy, M. Fletcher, G. Barnes, S.R. Durham and A. Sheikh. 2005. Ethnic variations in UK asthma frequency, morbidity, and health-service use: a systematic review and meta-analysis. Lancet 365: 312–17.
Socio-Economic Deprivation	Prescribing trends in a deprived area of East London was shown to lag about 2 years behind England as a whole	Naish, J., S. Eldridge, K. Moser and P. Sturdy. 2002. Did the London Initiative Zone Investment Programme affect general practice structure and performance in East London? A time series analysis of cervical screening coverage and asthma prescribing. Public Health 116: 361–367.
Rurality	Children in rural areas were less likely (35.7% v. 72.3%) to receive education on using PEF monitors and more likely to be undertreated with ICS (81.8% v.25.7%) compared to suburban and urban children, respectively	Reeves, M.J., S.R. Bohm, S.J. Korzeniewski and M.D. Brown. 2006. Asthma Care and Management Before an Emergency Department visit in children in Western Michigan: How Well does Care Adhere to Guidelines? Pediatrics 117: S118–S126.
ICS = Inhaled Corticosteroids; OR = Odds Ratio		

Second, following guidelines may have been inappropriate in some patients or during certain consultations. GPs may have abandoned or postponed some guideline criteria due to co-morbidities or to address the factors which the patient deemed most important to them during that consultation, respectively. This could have resulted in paradoxically poor performance against guidelines but high rankings of quality of care from patients.

Third, guidelines may have conflicted with patient autonomy. Though the correct treatment may have been offered, it is a patient's prerogative to believe that treatment based on guidelines is not in their best interests despite evidence, education, and encouragement to the contrary. Similarly, they may have sought care elsewhere including emergency departments—an outcome often labeled as a failure of PC. Such phenomena may not have been captured by these, particularly quantitative, studies.

PRACTICE & PROCEDURES

Examples of guidelines and how quality can be measured are presented in Tables 1 and 4, respectively.

KEY FACTS

- The vast majority (95%) of asthma healthcare takes place in PC.
- Clinical guidelines set standards for high quality asthma care.

Studies since 2002 have reported:

- Over half of asthma patients presenting to PC for any reason have been found to have uncontrolled asthma. Less than one-half of PC asthma patients received spirometry.
- Assessment of inhaler technique varied from 14 to 68%.
- Only about one-quarter of asthma patients received PAAPs.

SUMMARY POINTS

- Where it exists and is accessible, PC systems vary greatly between countries.
- In practice, healthcare quality is measured against clinical guidelines via audit.
- Recent studies highlight the need for continued quality of asthma care in PC.
- This can be improved with medical record systems tailored to asthma, prompts for annual review, and incentive schemes.
- Quality criteria in guidelines may differ from patient views on quality healthcare and may not take into account the demands of PC.

DEFINITIONS AND EXPLANATIONS

Benchmarking: Comparing one's performance against quality criteria to that of others. This can be done anonymously on the basis of individual GPs or by comparing practices against each other.

Clinical Audit: A cyclical process of measurement against standards, practice modification and re-measurement to confirm improvement.

Control: The degree to which symptoms or sequelae of asthma have been ameliorated by treatment and prevention strategies. This is not a static dichotomous concept, but a continuum over a time period generally of several weeks.

Exacerbation: Deterioration in the patient's previous status that is more than minor transient loss of control leading to only minor symptoms.

Peak Expiratory Flow (PEF): A commonly used measure of lung function.

Over/Underdosing: Prescribing the correct medication, but at an inappropriate dose.

Underprescribing: Failure to prescribe the correct medication based on the step-wise approach to asthma management.

LIST OF ABBREVIATIONS

GP	:	General Practitioner; in our chapter, any primary care doctor
ICS	:	Inhaled cortico-steroid(s)
IOM	:	Institute of Medicine
OECD	:	Organisation for Economic Cooperation and Development
PAAP	:	Personal Asthma Action Plan (written)
PC	:	Primary Care
PEF(R)	:	Peak Expiratory Flow (Rate)
pMDI	:	Pressurized Metered Dose Inhaler(s)
QOF	:	Quality and Outcomes Framework, a pay for performance scheme in the United Kingdom

ACKNOWLEDGEMENTS

Suzanne Hurd & Dr. Mark Levy from GINA.
Dr. Stephen D. Robinson from the University of East Anglia.

REFERENCES

Abudahish, A. and H. Bella. 2010. Adherence of primary care physicians in Aseer regions, Saudia Arabia to the National Protocol for the Management of Asthma. East Mediterr Health J 16: 171–175.

Aït-Khaled, N., D.A. Enarson, N. Bencharif, F. Bouladhib, L.M. Camara, E. Dagli, K. Djankine, B. Keita, B. Koadag, K. Ngoran, J. Odhiambo, S.E. Ottmani, D.L. Pham, O. Sow, M. Yousser and N. Zidouni. 2006. Implementation of asthma guidelines in health centres of several developing countries. Int J Tuberc Lung Dis 10: 104–109.

Aït-Khaled, N., D.A. Enarson, C.Y. Chiang, G. Marks and K. Bissell. Management of asthma: a guide to the essentials of good clinical practice. 2008. International Union Against Tuberculosis and Lung Disease. Paris, France.

Arbuthnott, A. and D. Sharpe. 2009. The effect of physician-patient collaboration on patient adherence in non-psychiatric medicine. Patient Educ Couns 77: 60–67.

Barnes, P.J. 2004. Asthma guidelines: recommendations versus reality. Respir Med 98: S1–S7.

Bateman, E., C. Feldman, R. Mash, L. Fairall, R. English and A. Jithoo. 2009. Systems for the management of respiratory disease in primary care—an international series: south Africa. Prim Care Respir J 18: 69–75.

Burgess, R. 2011. New Principles of Best Practice in Clinical Audit. Healthcare Quality Improvement Partnership, Radcliff, Oxford UK, ISBN 978 184619 221 0

Byrnes, P., C. McGoldrick and M. Crawford. 2010. Asthma cycle of Care Attendance: Overcoming therapeutic inertia using an asthma clinic. Aust Fam Physician 39: 318–320.

Campbell, S.J., J. Braspenning, A. Hutchinson and M. Marshall. 2004. Research methods used in developing and applying quality indicators in primary care *In:* Grol et al Quality Improvement Research, BMJ Publishing Group 2004.

Campbell, S.J., D. Reeves, E. Kontopantelis, E. Middleton, B. Sibbald and M. Roland. 2007. Quality of Primary Care in England with the Introduction of Pay for Performance. N Engl J Med 357: 181–190.

Carlton, G., D.O. Luca, E.F. Ellis, K. Conboy-Ellis, O. Shoheiber and D.A. Stempel. 2005. The Status of Asthma Control and Asthma Prescribing Practices in the United States: Results of a Large Prospective Asthma Control Survey of Primary Care Practices J Asthma 42: 529–535.

Chapman, K.R., L.P. Boulet, R.M. Rea and E. Franssen. 2008. Suboptimal asthma control: prevalence, detection and consequences in general practice. Eur Respir J 31: 320–325.

Dixon-Woods, M., A. Anwar, B. Young and A. Brooke. 2002. Lay evaluation of services for childhood asthma. Health Soc. Care Community 10: 503–511.

Donabedian, A. 1980. Explorations in Quality Assessment and Monitoring. Vol. 1. The Definition of Quality and Approaches to its Assessment. Health Administration Press. Ann Arbor, MI, USA.

Douglass, J., R. Aroni, D. Goeman, K. Stewart, S. Sawyer, F. Thien and M. Abramson. 2002. A qualitative study of action plans for asthma. BMJ 324: 1003–1006.

Gamble, J., M. Stevenson, E. McClean and L.G. Heaney. 2009. The Prevalence of Nonadherence in Difficult Asthma. Am J Respir Crit Care Med 180: 817–822.

Gibson, P.G., A. Wilson, M.J. Abramson, P. Haywood, A. Bauman, M.J. Hensley, E.H. Walters and J.J.L. Roberts. 2009. Self-management education and regular practitioner review for adults with asthma (Review) Cochrane Database Syst Rev 2002, 3. John Wiley & Sons, Ltd.

[GINA] Global Initiative for Asthma. 2009. Global Strategy for Asthma Management and Prevention. Medical Communications Resources, Inc. www.ginasthma.org accessed on 31 January 2010.

Goeman, D.P., C.D. Hogan, R.A. Aroni, M.J. Abramson, S.M. Sawyer, K. Stewart, L.A. Sanci and J.A. Douglass. 2005. Barriers to delivering asthma care: a qualitative study of general practitioners. Med J Aust 183: 457–460.

Hasselgren, M., D. Gustafsson, B. Ställberg, K. Lisspers and G. Johansson. 2005. Management, asthma control and quality of life in Swedish adolescents with asthma. Acta Paediatr 94: 682–688.

Hoskins, G., C. McGowan, P.T. Donnan, J.A.R. Friend and L. Osman. 2005. Results of a national asthma campaign survey of primary care in Scotland. Int J Qual Health Care 17: 209–215.

Howie, J.G.R., D. Heaney and M. Maxwell. 2004. Quality, core values and the general practice consultation: issues of definition, measurement and delivery. Fam Pract 21: 458–468.

[IOM] Institute of Medicine. 2001. Crossing the Quality Chasm. National Academy Press, Washington, DC, USA.

Jadad, A.R., M. Moher, G.P. Browman, L. Booker, C. Sigouin, M. Fuentes and R. Stevens. 2000. Systematic reviews and meta-analyses on threatment of asthma: critical evaluation. BMJ 320: 537–540.

Kelley, E. and J. Hurst. 2006. Health Care Quality Indicators Project Conceptual Framework Paper. OECD Health Working Papers 23. http://www.oecd.org/dataoecd/1/36/36262363.pdf

Kringos, D.S., W.G.W. Boerma, A. Hutchinson, J. van der Zee and P.J. Groenewegen. 2010. The breadth of primary care: a systematic literature review of its core dimensions. BMC Health Serv Res 10: 65.

Lieu, T.A., J.A. Finkelstein, P. Lozano, A.M. Capra, F.W. Chi, N. Jensvold, C.P. Quesenberry and H.J. Farber. 2004. Cultural competence Policies and Other Predictors of Asthma Care quality for Medicaid-Insured Children. Pediatrics 114: pe102–e110.

Lyte, G., L. Milnes, P. Keating and A. Finke. 2007. Review management for children with asthma in primary care: a qualitative case study. J Nurs Healthc. Chronic Illn 16: 123–132.

Maxwell, R.J. 1992. Dimensions of quality revisited: from thought to action. Qual Health Care 1: 171–7.

Mintz, M., A.W. Gilsenan, C.L. Bui, R. Ziemiecki, R.H. Stanford, W. Lincourt and H. Ortega. 2009. Assessment of asthma control in primary care. Curr Med Res Opin 25: 2523–2531.

Moth, G., P.O. Schiotz and P. Vedsted. 2008. A Danish population-based cohort study of newly diagnosed asthmatic children's care pathway—adherence to guidelines. BMC Health Serv Res 8: 130.

Neville, R.G., G. Hoskins, C. McCowan and B. Smith. 2004. Pragmatic 'real world' study of the effect of audit of asthma on clinical outcome. Prim. Care Respir J 13: 198–204.

Nizami, T.A. and B. Mash. 2005. Inhaled steroid use in adult asthmatics – experience at a primary health care centre. S Afr Med J 95: 169–170.

Powell H. and P.G. Gibson. 2009. Options for self-management education for adults with asthma (review). Cochrane Database Syst Rev John Wiley & Sons, Ltd. 11 March 2002.

Ring, N., C. Malcolm, S. Wyke, S. MacGillivray, D. Dixon, G. Hoskins, H. Pinnock and A. Sheikh. 2007. Promoting the use of Personal Asthma Action Plans: A systematic review. Prim Care Respir J 16: 271–283.

Sans-Corrales, M., E. Pujol-Ribera, J. Gené-Badia, M.I. Pasarín-Rua, B. Iglesias-Pérez and J. Casajuana-Brunet. 2006. Family medicine attributes related to satisfaction, health and costs. Fam Pract 23: 308–316.

Schneider, A., K. Biessecker, R. Quinzler, P. Kaufmann-Kolle, F. J. Meyer, M. Wensing and J. Szecsenyi. 2007. Asthma patients with low perceived burden of illness: a challenge for guideline adherence. J Eval Clin Pract 13: 846–852.

Silver, A., J. Figge, D.L. Haskin, V. Pryor, K. Fuller, T. Lemme, N. Li and M.J O'Brien. 2011 (in press). An asthma and diabetes Quality Improvement Project: Enhancing Care in Clinics and Community Health Centers. J Community Health 29 July.

Starfield, B. 2009. Primary Care and Equity in Health: The Importance to Effectiveness and Equity or Responsiveness. Humanity Soc 33: 56–73.

Steel, N., D. Meltzer and I. Lang. 2007a. Improving Quality in Gillam, S. and J. Yates and P. Badrinath. Essential Public Health: Theory and Practice. Cambridge University Press. ISBN: 9780521689830

Steel, N., S. Maisey, A. Clark, R. Fleetcroft and A. Howe. 2007b. Quality of clinical primary care and targeted incentive payments: an observational study. Br J Gen Pract 57: 449–454.

To, T., A. Guttmann, M.D. Lougheed, A.S. Gershon, S.D. Dell, M.B. Stanbrook, C.Wang, S. McLimont, J. Vasilevska-ristovska, E.J. Crighton and D.N Fishman. 2010. Evidence-based performace indicators of primary care for asthma: a modified RAND Appropriateness Method. Int J Qual Health Care 22: 476–485.

Weidinger, P., J.L.G. Nilsson and U. Lindblad. 2009. Adherence to diagnostic guidelines and quality indicators in asthma and COPD in Swedish primary care. Pharmacoepidemiol. Drug Saf 18: 393–400.

Wiener-Ogilvie, S., H. Pinnock, G. Huby, A. Sheikh, M.R. Partridge and J. Gillies. 2007. Do practices comply with key recommendations of the British Asthma Guideline? If not, why not? Prim Care Respir J 16: 369–377.

2

Prostaglandins and Leukotrienes: Mediators of Inflammation in Asthma

Neil L. Misso,[1,a,*] *Shashi Aggarwal*[1,b] and *Philip J. Thompson*[1,c]

ABSTRACT

Prostaglandins and leukotrienes are key mediators of inflammation and remodelling of the airways in asthma. These lipid mediators are synthesized by the activities of cyclooxygenase and 5-lipoxygenase enzymes, respectively, on arachidonic acid liberated from cell membranes. Downstream enzymes catalyse the formation of prostaglandin (PG)D_2, PGE_2 and the cysteinyl leukotrienes (cysLT), LTC_4, LTD_4 and LTE_4 that are implicated in diverse biological effects related to the pathophysiology of asthma. In the lungs, the main sources of prostaglandins are airway epithelium and macrophages (PGE_2), as well as mast cells (PGD_2), whereas the cysLT are mainly produced by eosinophils and mast cells. Prostaglandins and leukotrienes exert their various biological activities by binding to a number of seven transmembrane, G protein coupled receptors. PGE_2 has

[1]Lung Institute of Western Australia, Centre for Asthma, Allergy & Respiratory Research, The University of Western Australia, Ground Floor E Block, Sir Charles Gairdner Hospital, Nedlands, WA 6009, Australia;
[a]Email: nmisso@liwa.uwa.edu.au
[b]Email: shashi@liwa.uwa.edu.au
[c]Email: pjthomps@liwa.uwa.edu.au
List of abbreviations after the text.
*Corresponding author

smooth muscle relaxant and anti-inflammatory effects in the lungs of asthmatic patients, whereas PGD_2 is a potent bronchoconstrictor and also causes vasodilatation and the recruitment and activation of Th2 lymphocytes and eosinophils. CysLT are likely to be key mediators causing bronchoconstriction in asthmatic patients, and are also potent mediators of inflammation and airway remodelling. The balance between the production of cysLT and PGE_2 in the airways may be a determinant of chronic asthma severity, as well as airway inflammation and remodelling. $CysLT_1$ receptor antagonists are widely used in the treatment of asthmatic patients; however, clinical responsiveness to these drugs varies greatly and further research is required into the factors that might influence responsiveness to anti-leukotriene therapy. Therapeutic strategies that more specifically target the effects of prostaglandins in asthma will have to await the development of prostaglandin receptor antagonists and synthase inhibitors.

INTRODUCTION

Asthma is a common disease of the airways that involves episodes of bronchoconstriction with chronic inflammation and eventual structural remodelling of the airways. Asthmatic airway inflammation is characterized by the infiltration of eosinophils, neutrophils, mast cells and activated T helper 2 (Th2) lymphocytes, which together with structural cells such as airway epithelial cells, smooth muscle cells, endothelial cells and fibroblasts, release a variety of inflammatory mediators that are responsible for many of the pathophysiological processes occurring in the airways, including bronchoconstriction, mucosal and submucosal oedema, increased secretion of mucus and remodelling of the airways (Hamid and Tulic 2009).

A large number of mediators have been implicated in the pathophysiological changes that occur in asthma (Barnes et al. 1998). These include pro-inflammatory and anti-inflammatory cytokines, growth factors, chemokines, peptide mediators, amines and prostaglandins and leukotrienes, which are the focus of this chapter.

Since the 1970s there has been intensive research into the mechanisms regulating the synthesis and metabolism of prostaglandins and leukotrienes, the receptors and signalling pathways through which they mediate their biological effects, and their precise roles in asthma and airway inflammation. More recently, there has been progress in the development and clinical application of drugs that inhibit the synthesis or biological effects of prostaglandins and leukotrienes.

SYNTHESIS OF PROSTAGLANDINS AND LEUKOTRIENES

In the lungs, prostaglandins are mainly produced by the airway epithelium, macrophages and mast cells, whereas the main sources of leukotrienes are mast cells and eosinophils. All prostaglandins and leukotrienes are derived from the 20-carbon polyunsaturated fatty acid, arachidonic (eicosatetraenoic) acid, and are therefore also referred to as eicosanoids (Fig. 1). Arachidonic acid is a ubiquitous component of cell membranes, and the first step in its conversion to prostaglandins and leukotrienes is catalysed by cytosolic and secretory phospholipase A_2 enzymes that liberate arachidonic acid from membrane phospholipids.

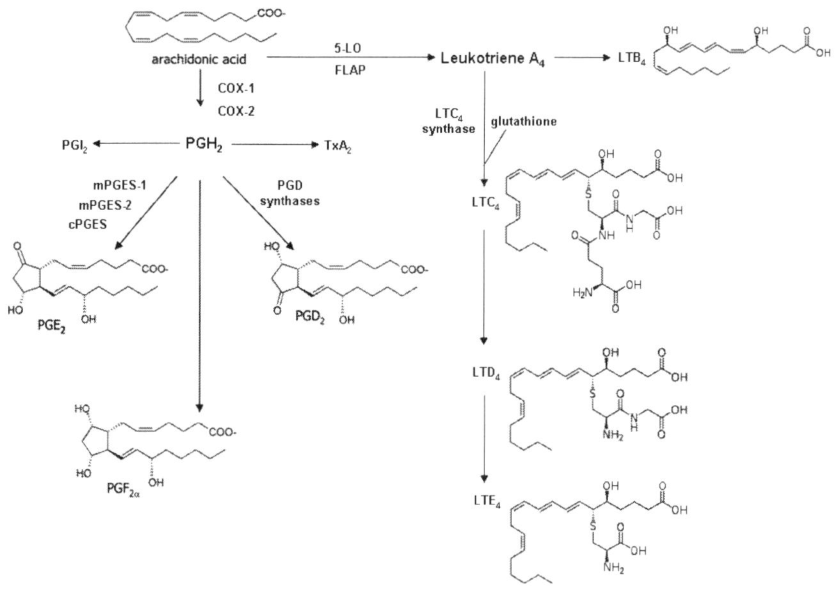

Figure 1 Synthesis of prostaglandins and leukotrienes from arachidonic acid. COX, cyclooxygenase; LT, leukotriene; 5-LO, 5-lipoxygenase; FLAP, 5-lipoxygenase activating protein; PG, prostaglandin; mPGES, microsomal PGE synthase; cPGES, cytosolic PGE synthase; TxA_2, thromboxane A_2.

Synthesis of Prostaglandins

Arachidonic acid is the substrate for two cyclooxygenase enzymes (COX-1 and COX-2) that catalyse the formation of the intermediate, prostaglandin (PG)H_2. COX-1 is expressed constitutively in most cells, whereas expression of COX-2 is generally low but is rapidly up-regulated when cells receive an appropriate stimulus. The early prostaglandin response may depend on COX-1 activity, whereas subsequent amplification of prostaglandin production, with progression of lung inflammation, is probably mediated by COX-2 (Petrovic et al. 2006).

PGH_2 is converted to the biologically active prostaglandins, PGD_2, PGE_2, $PGF_{2\alpha}$ and PGI_2, as well as thromboxane (Tx)A_2, by specific synthase enzymes (Tilley et al. 2001). The differential expression of these synthases in specific cell types determines the overall profile of prostaglandin production within a tissue. In addition, the profile of prostaglandins produced by a particular cell type may be altered by specific inflammatory stimuli and by coupling of COX-1 or COX-2 with specific synthases. Macrophages and airway epithelium express three isoforms of PGE synthase, one of which is cytosolic (cPGES), while two are membrane bound (mPGES-1, mPGES-2) (Fig. 1). cPGES is constitutively expressed and catalyses the isomerization of PGH_2 produced by COX-1 (Samuelsson et al. 2007). mPGES-1 is an inducible isoform that acts on PGH_2 produced by COX-2, and its up-regulation in response to inflammatory stimuli is coordinated with that of COX-2. Two types of PGD synthase have been identified; a lipocalin type and a haematopoietic type (Urade and Eguchi 2002). Mast cells express haematopoietic PGD synthase and rapidly secrete PGD_2 when stimulated by antigens.

Synthesis of Leukotrienes

The enzyme 5-lipoxygenase (5-LO) cooperates with 5-lipoxygenase-activating protein (FLAP) to catalyse the conversion of arachidonic acid to leukotriene (LT)A_4 (Fig. 1). This intermediate is converted to LTB_4 by LTA_4 hydrolase, or to LTC_4 by conjugation with reduced glutathione, a reaction catalysed by LTC_4 synthase (Peters-Golden and Henderson 2007). LTC_4 is further metabolized to LTD_4 by hydrolysis of glutamic acid and to LTE_4 by hydrolysis of both glutamic acid and glycine from the glutathione moiety.

5-LO shuttles between the cytoplasm and the nuclear membrane, where it cooperates with FLAP and LTC_4 synthase, which are members of the super family of membrane-associated proteins in eicosanoid and glutathione metabolism (MAPEG), in the production of LTC_4. Recent studies have increased our understanding of the compartmentalization, crystal structures and catalytic mechanisms of these enzymes (Newcomer and Gilbert 2010). Other factors regulating the capacity for leukotriene production include the amount of free arachidonic acid in the cell, the level of expression and catalytic activities of the 5-LO and LTC_4 synthase enzymes, and the modulation of 5-LO activity by ATP, nitric oxide and reactive oxygen species (Peters-Golden and Henderson 2007).

PROSTAGLANDIN AND LEUKOTRIENE RECEPTORS

Prostaglandins and leukotrienes exert their biological activities by binding to a number of seven transmembrane G protein-coupled receptors. These receptors are located on the plasma membrane of structural and inflammatory cells, and differential expression of these receptors in various cells and tissues determines the functional activity of prostaglandins and leukotrienes in health and disease.

Prostaglandin Receptors

The prostaglandin receptor family comprises nine members that each show specificity for one class of prostanoid. These are the EP1, EP2, EP3 and EP4 receptors for PGE_2, the DP and CRTH2 receptors for PGD_2, the TP receptor for TxA_2, the IP receptor for prostacyclin, and the FP receptor for $PGF_{2\alpha}$ (Hata and Breyer 2004). Although all the prostanoid receptors are coupled to heterotrimeric G proteins, their function in physiological and pathological settings is determined by complex interactions arising from differences in ligand affinities, receptor expression and coupling to intracellular signal transduction pathways (Table 1). Thus binding of a prostaglandin to its specific receptor may produce varying biological effects in different cells and tissues.

Table 1 G protein coupling and signal transduction pathways for the prostaglandin receptors.

Ligand	Receptor subtype	G protein coupling	Signal transduction
TxA_2	TP	G_q, G_s (α), G_i (β) G_h (α), G_{12}	↑IP_3/DAG/Ca^{2+},↑cAMP ↓cAMP
PGI_2	IP	G_s, G_q, G_i	↑cAMP,↑IP_3/DAG,↓cAMP
$PGF_{2\alpha}$	FP	G_q	↑IP_3/DAG
PGE_2	EP1		↑Ca^{2+}
	EP2	G_s	↑cAMP
	EP3	G_i, G_q, G_s	↓ cAMP,↑ IP_3/DAG,↓cAMP
	EP4	G_s	↑ cAMP
PGD_2	DP	G_s	↑cAMP,↑Ca^{2+}
	CRTH2	G_i	↓cAMP,↑Ca^{2+}

PG, prostaglandin; TxA_2, thromboxane A_2; IP_3, inositol 1,4,5-trisphosphate; DAG, diacylglycerol; cAMP, cyclic adenosine monophosphate; ↑ increase; ↓ decrease.

PGD_2 Receptors

The pro-inflammatory effects of PGD_2 are mediated by both DP and CRTH2 receptors. Activation of the DP receptor increases intracellular cAMP (Table 1) and inhibits the function of T cells and other inflammatory cells (Tilley et al. 2001). However, the DP receptor is also expressed on bronchial epithelium, where it mediates the production of chemokines and cytokines involved in the recruitment of lymphocytes and eosinophils (Kabashima and Narumiya 2003). Airway hyperreactivity and Th2-mediated lung inflammation were greatly reduced in mice that were deficient in the DP receptor, suggesting that it may mediate biological effects similar to those observed in asthmatic patients (Matsuoka et al. 2000). Furthermore, variation in the DP receptor gene is associated with susceptibility to asthma (Oguma et al. 2004).

The CRTH2 receptor is expressed on Th2 lymphocytes, eosinophils and basophils, and binding of PGD_2 to the CRTH2 receptor stimulates chemotaxis and recruitment of these cells through a decrease in cAMP and an increase in intracellular Ca^{2+} (Hata and Breyer 2004) (Table 1). However, PGD_2 also inhibits the apoptosis of eosinophils through DP receptors (Gervais et al. 2001). Therefore, the contrasting actions of PGD_2 at the DP and CRTH2 receptors may contribute to eosinophilic infiltration of the airways in allergic asthma.

PGE_2 Receptors

The diverse effects of PGE_2 are mediated by four receptors that show differential coupling to G proteins and intracellular signal transduction pathways (Table 1) (Hata and Breyer 2004). EP1 receptors mediate smooth muscle contraction through an increase in intracellular Ca^{2+} that appears to be independent of the G_q protein; however, EP1 receptor signal transduction pathways are not well understood. Both the EP2 and EP4 receptors are coupled to a G_s protein, thereby increasing cAMP levels and having a relaxant effect on smooth muscle. However, the increase in cAMP is much less for EP4 than for EP2, whereas stimulation of EP4, but not EP2, results in phosphatidyl 3-kinase-dependent phosphorylation of extracellular signal related kinase 1/2, suggesting that there are important functional differences and unique roles for these two receptors (Hata and Breyer 2004). The EP3 receptor has multiple splice variants due to alternative splicing of its C-terminal tail. At least eight EP3 splice variants have been identified in humans and these differ in their constitutive activity, signal transduction pathways, desensitization and intracellular trafficking (Hata and Breyer 2004). Activation of the EP3 receptor decreases intracellular cAMP, although activation of some of its splice variants appears to stimulate the generation of cAMP and inositol 1,4,5-trisphosphate (IP_3) (Table 1).

The four EP receptors are differentially expressed on T and B cells, dendritic cells, macrophages, neutrophils and eosinophils, suggesting that PGE_2 may have multiple roles in the modulation of immune responses (Tilley et al. 2001) (Fig. 2). Binding of PGE_2 to EP2 and EP4 receptors inhibits the proliferation of lymphocytes, as well as cytokine production by lymphocytes and macrophages (Tilley et al. 2001, Hata and Breyer 2004). PGE_2 also inhibited eosinophil

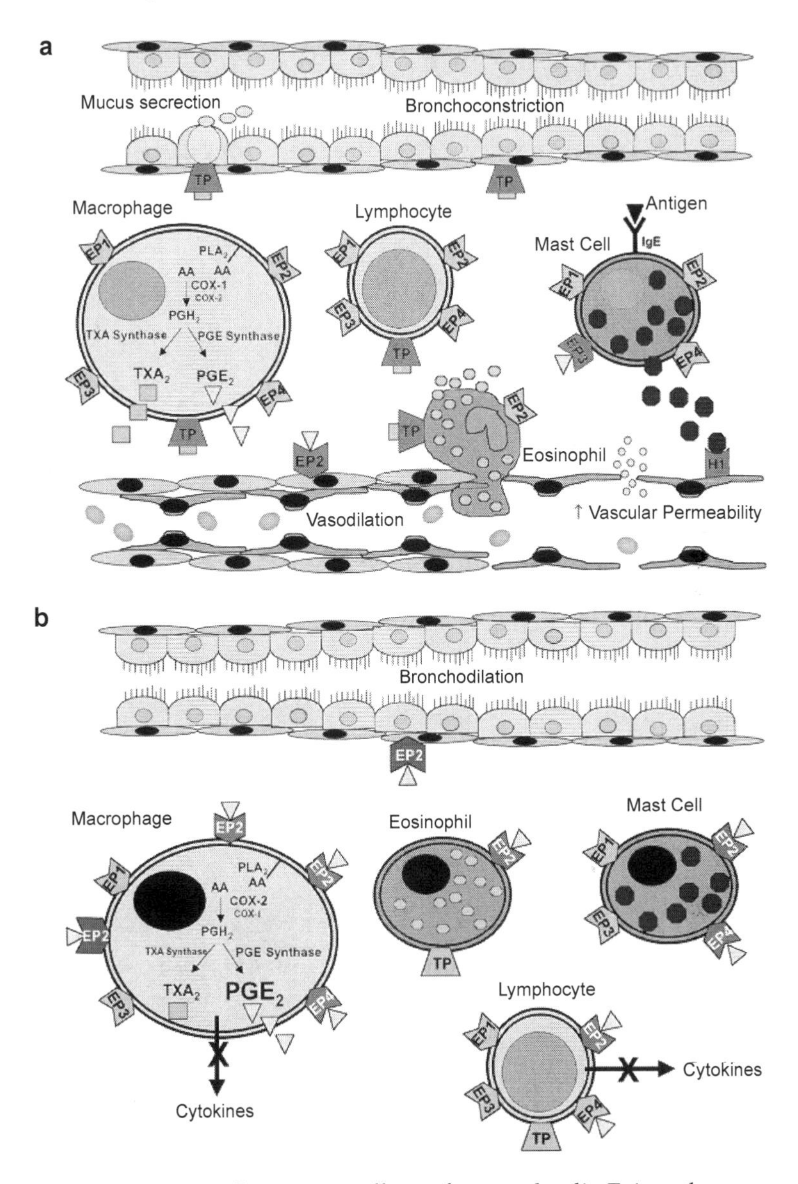

Figure 2 Pro- and anti-inflammatory effects of prostaglandin E_2 in asthma. (a) PGE_2 causes vasodilatation by activating EP2 receptors on vascular smooth muscle, and enhances the release of mediators from mast cells. (b) As a consequence of airway inflammation, PGE_2 synthesis by macrophages and airway epithelium is increased. PGE_2 inhibits lymphocyte activation and causes bronchodilatation by activating EP2 receptors on airway smooth muscle. Modified from Tilley et al. 2001, and reproduced with permission from the American Society for Clinical Investigation © 2001.

apoptosis through a mechanism involving EP2 receptors and an increase in cAMP (Peacock et al. 1999) (Figs. 3 and 4). Therefore the binding of PGE_2 to the EP receptors on immune cells may have both pro- and anti-inflammatory consequences, depending on the cell type, the receptor subtypes expressed and the pathophysiological context (Fig. 2).

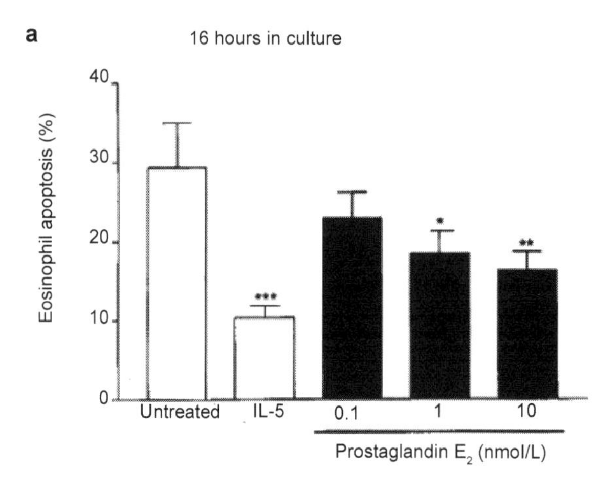

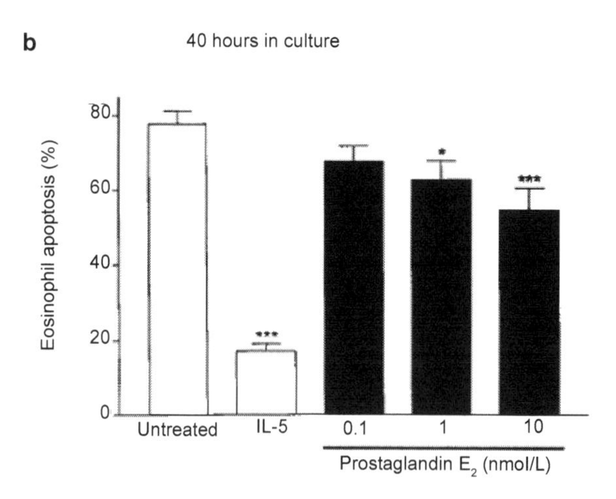

Figure 3 Prostaglandin E_2 inhibits the apoptosis of eosinophils.
Purified human eosinophils were cultured for 16 hr (a) or 40 hr (b) in the presence of prostaglandin E_2 or IL-5 (100 U/mL) and apoptosis was assessed by flow cytometry, as the percentage of cells binding annexin V. Bars indicate means and SEM. *$P < 0.05$, **$P < 0.01$, ***$P < 0.001$, compared with untreated eosinophils. Reproduced from Peacock et al. 1999, with permission from Elsevier Ltd. © 1999.

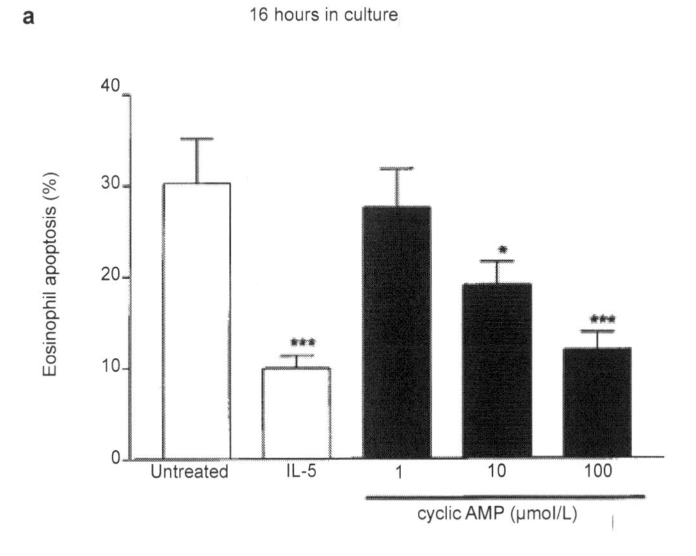

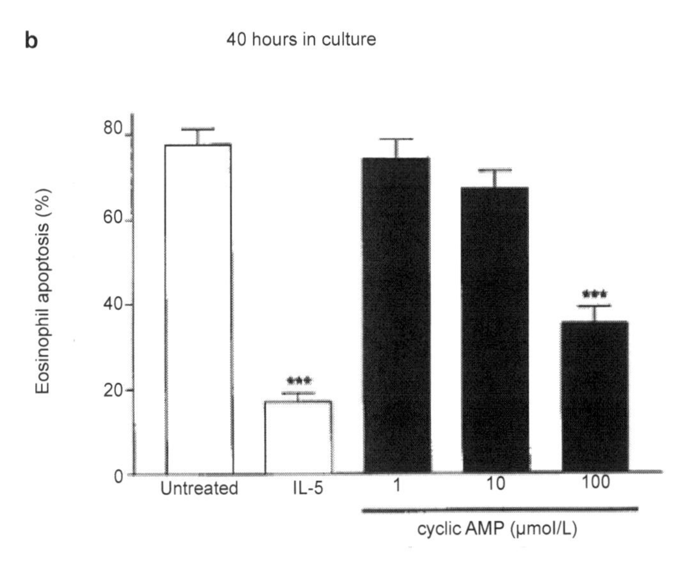

Figure 4 Cyclic adenosine monophosphate (cAMP) inhibits the apoptosis of eosinophils.
Purified human eosinophils were cultured for 16 hr (a) or 40 hr (b) in the presence of dibutyryl cAMP or IL-5 (100 U/mL) and apoptosis was assessed by flow cytometry, as the percentage of cells binding annexin V. Bars indicate means and SEM. $^{*}P < 0.05$, $^{***}P < 0.001$, compared with untreated eosinophils. Reproduced from Peacock et al. 1999, with permission from Elsevier Ltd. © 1999.

Cysteinyl Leukotriene Receptors

LTC_4, LTD_4 and LTE_4 exert their biological activities by binding to two G protein-coupled receptors, termed $CysLT_1$ and $CysLT_2$. The $CysLT_1$ receptor is highly expressed in the spleen and in eosinophils, neutrophils, mast cells, macrophages, B lymphocytes and plasma cells, with weaker expression in the lungs, small intestine, pancreas and placenta. Expression of $CysLT_1$ receptor mRNA and protein was significantly greater in patients with stable asthma and those experiencing acute exacerbations, compared with healthy control subjects (Zhu et al. 2005). The $CysLT_2$ receptor protein shares only 38% homology with the $CysLT_1$ receptor (Singh et al. 2010). $CysLT_2$ is highly expressed in spleen, heart, adrenal gland, brain and peripheral blood leukocytes. The strong expression of $CysLT_2$ in eosinophils, suggests the possibility of as yet unidentified roles in these cells (Singh et al. 2010). However, expression of $CysLT_2$ in airway smooth muscle is relatively weak, compared with expression of $CysLT_1$.

High affinity binding of the LT ligands results in activation of a G_q protein, hydrolysis of guanosine triphosphate (GTP), phospholipase C mediated generation of IP_3 and diacylglycerol, and increases in intracellular Ca^{2+} (Singh et al. 2010). For $CysLT_1$ the order of potency is $LTD_4 >> LTC_4 > LTE_4$, whereas for $CysLT_2$ it is $LTD_4 = LTC_4 > LTE_4$. The weak affinities of $CysLT_1$ and $CysLT_2$ for LTE_4 contrast with its relative potency for enhancing airway responsiveness to histamine and increasing eosinophil recruitment and vascular permeability, suggesting the probable existence of additional CysLT receptor subtypes (Lee et al. 2009). Furthermore the airways of asthmatic patients demonstrate marked selective hyperresponsiveness to LTE_4 (Austen et al. 2009). Recent studies suggest the existence of receptors that are more selective for LTE_4, including the $P2Y_{12}$ purinergic receptor and the $CysLT_E$ receptor, which was functional in the skin of mice lacking $CysLT_1$ and $CysLT_2$. The $P2Y_{12}$ purinergic receptor may mediate pro-inflammatory effects of LTE_4, and may therefore be a potential therapeutic target in asthma (Parachuri et al. 2009).

The expression of the $CysLT_1$ and $CysLT_2$ receptors is regulated by inflammatory cytokines. Thus IL-5 increased the expression of $CysLT_1$ by a transcriptional mechanism, and IL-4 and IL-3, but not interferon-γ, induced $CysLT_1$ expression in monocytes and macrophages (Singh et al. 2010). In contrast, interferon-γ induced both $CysLT_1$ and $CysLT_2$ expression in airway smooth muscle cells.

$CysLT_2$ mRNA expression in endothelial cells was up regulated by IL-4; however, this effect was inhibited by tumour necrosis factor-α and IL-1β. In addition, interferon-γ was involved in the up regulation of $CysLT_2$ receptors on eosinophils during asthma exacerbations (Fujii et al. 2005). These observations suggest that Th1 and Th2 cytokines regulate the expression of CysLT receptors, with potentially important pathological consequences in asthma.

PATHOPHYSIOLOGICAL ROLE OF PROSTAGLANDINS AND LEUKOTRIENES IN ASTHMA

The importance of prostaglandins and leukotrienes in the pathophysiology of asthma has prompted research into the development of specific inhibitors and receptor antagonists that might improve asthma management, reduce the high rates of morbidity associated with the disease, and avoid the side effects associated with corticosteroids, which are the current mainstay of preventive treatment. One of the most important achievements in asthma therapeutics has been the development and translation into clinical practice of the leukotriene receptor antagonists. Although research into specific prostaglandin receptor antagonists is at an early stage, the next decade may see the development and clinical application of antagonists that specifically target the numerous effects of prostaglandins in asthma.

Effects of Prostaglandins in Asthma

Prostaglandin E_2

PGE_2 relaxes human airway smooth muscle *in vitro* (Knight et al. 1995), an effect that is mediated by EP2 receptors (Barnes et al. 1998). Inhaled PGE_2 also causes bronchodilatation in normal subjects and protects against exercise-, metabisulphite-, and allergen-induced bronchoconstriction in asthmatic patients (Barnes et al. 1998). In addition, inhalation of PGE_2 markedly reduces the recruitment of eosinophils into the lungs of asthmatic patients (Vancheri et al. 2004). *In vitro* studies have shown that PGE_2 inhibits the formation of leukotrienes, TxA_2 and PGD_2 by mast cells and eosinophils, and also down-regulates the expression of adhesion molecules in airway smooth muscle and endothelial cells, thereby inhibiting the transendothelial migration of T lymphocytes and Th2 differentiation (Vancheri et al. 2004). In addition, PGE_2 modulates tissue repair and fibrosis in the lungs by stimulating the production of the anti-inflammatory cytokine, IL-10, promoting epithelial growth, and inhibiting collagen synthesis and the transformation of myofibroblasts to fibroblasts.

Overall, PGE_2 limits immune-inflammatory processes and has a beneficial effect in the lungs of asthmatic patients (Tilley et al. 2001; Vancheri et al. 2004) (Fig. 2). This is supported by recent observations that PGE_2 concentrations were increased in induced sputum of patients with more severe asthma, possibly as a bronchoprotective, anti-inflammatory response to counter the effects of cysLT and other pro-inflammatory mediators in the airways (Aggarwal et al. 2010) (Fig. 5). Interestingly, PGE_2 concentrations and the ratio of PGE_2 to cysLT in induced sputum were significantly lower in patients with

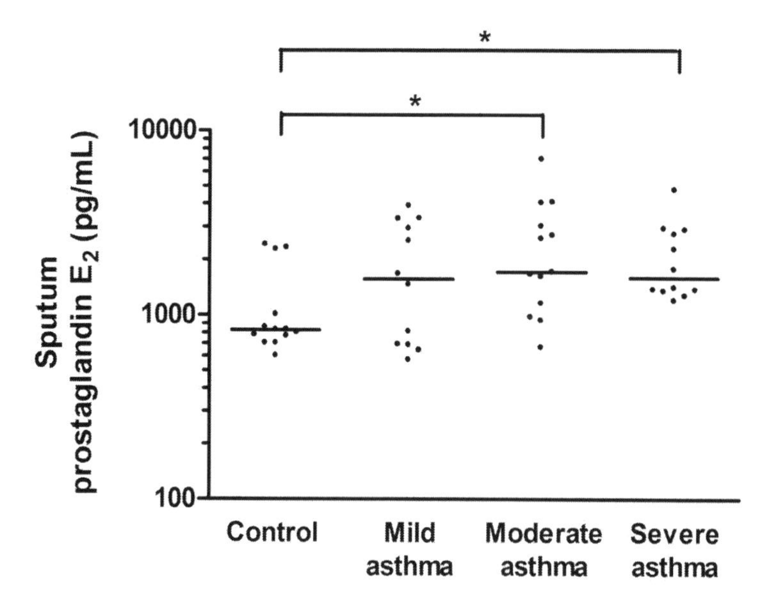

Figure 5 Prostaglandin E_2 concentrations are increased in induced sputum of patients with moderate and severe asthma.
Sputum prostaglandin E_2 concentrations in healthy control subjects (n = 13), and patients with mild (n =12), moderate (n = 13) or severe asthma (n = 12). Bars indicate median values. *$P < 0.05$. Reproduced from Aggarwal et al. 2010, with permission from John Wiley and Sons © 2010.

eosinophilic, as compared with non-eosinophilic inflammation of the airways (Fig. 6). These findings suggest that a deficiency in PGE_2 production relative to the formation of cysLT may adversely affect lung function and contribute to persistence of symptoms and airway remodelling in patients with eosinophilic airway inflammation.

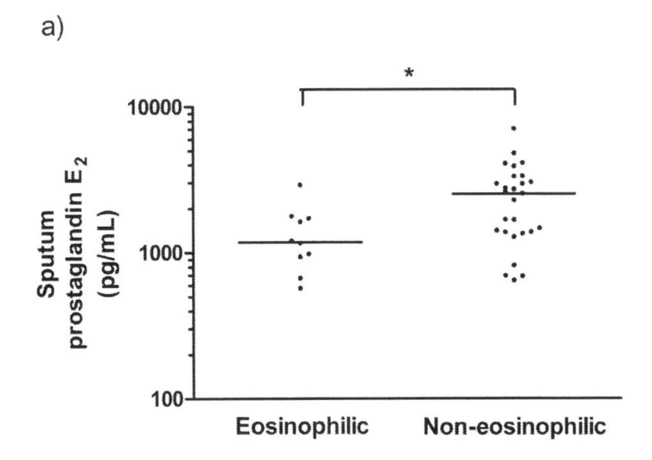

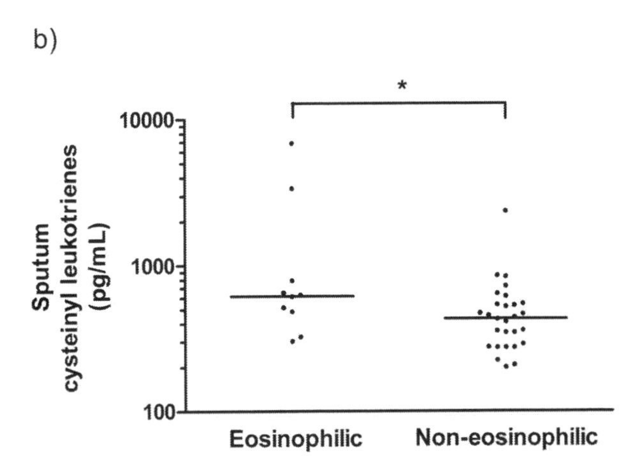

Figure 6 Prostaglandin E_2 concentrations are decreased, whereas cysteinyl leukotriene concentrations are increased in induced sputum of patients with eosinophilic asthma.

Prostaglandin E_2 (a) and cysteinyl leukotriene (b) concentrations in sputum of patients with eosinophilic (n = 10) or non-eosinophilic (n = 28) airway inflammation. Bars indicate median values. $^*P < 0.05$. Reproduced from Aggarwal et al. 2010, with permission from John Wiley and Sons © 2010.

Prostaglandin D_2

PGD_2 is a potent bronchoconstrictor, although this effect is mediated by the TP receptor, possibly through the formation of the PGD_2 metabolite, 9α,11β-prostaglandin F_2 (Pettipher et al. 2007). Binding of PGD_2 to DP receptors on bronchial smooth muscle results in bronchodilatation; however the predominant effect of PGD_2 in the airways is bronchoconstriction, indicating that bronchodilatation mediated by DP receptors is overcome by a more potent bronchoconstrictor effect of TP receptor activation.

PGD_2 has a number of other biological effects that are relevant to the pathophysiology of asthma. Thus, PGD_2 acts through the DP receptor to cause vasodilatation, leading to erythema and oedema (Pettipher et al. 2007). By interacting with the DP receptor on dendritic cells, PGD_2 favours an environment in which T lymphocytes are polarized towards a Th2 phenotype. In addition, PGD_2 acts through the CRTH2 receptor to potently stimulate the recruitment and activation of Th2 lymphocytes, and thereby the production of cytokines such as IL-4, IL-5 and IL-13 that mediate a number of pathological effects in allergic airway inflammation and asthma. Because of these wide-ranging and potent effects, the development of DP and CRTH2 antagonists has been identified as a potentially useful therapeutic strategy in asthma (Pettipher et al. 2007).

Effects of Cysteinyl Leukotrienes in Asthma

As potent bronchoconstrictors, cysLT play an important role in the pathophysiology of both acute and chronic asthma (Singh et al. 2010). LTD_4 induces contraction of human airway smooth muscle by binding to $CysLT_1$ receptors but the mechanism appears to be partly independent of increases in Ca^{2+} and may involve activation of a protein kinase C isoform (Holgate et al. 2003). Further studies are necessary to elucidate the precise mechanisms by which cysLT stimulate the contraction of airway smooth muscle and induce airway hyperresponsiveness in asthmatic subjects.

CysLT are potent mediators of inflammation and airway remodelling in asthma, and have been implicated in the trafficking and degranulation of eosinophils in the lungs, increased microvascular permeability leading to pulmonary oedema, and increased mucus secretion (Holgate et al. 2003). Airway remodelling, which may lead to irreversible airway obstruction, occurs as a result of enhanced proliferation of smooth muscle cells and fibroblasts, and increased deposition of matrix proteins such as collagen in the airway walls. CysLT have been implicated as playing a role in all these processes in the airways of asthmatic patients (Holgate et al. 2003).

Airway smooth muscle tone may be regulated by a balance between the bronchoconstrictor effects of cysLT and the relaxant effects of PGE_2 (Singh et al. 2010). Furthermore, PGE_2 inhibits, whereas cysLT promote fibroblast proliferation, synthesis of collagen, the recruitment of leukocytes and the generation of cytokines and growth factors (Holgate et al. 2003). Therefore, the balance between the production of cysLT and PGE_2 in the airways may be an important determinant not only of chronic asthma severity, but also of airway inflammation and remodelling in asthmatic patients (Aggarwal et al. 2010) (Fig. 6).

PRACTICE AND PROCEDURES

Measurement of Prostaglandins

Prostaglandins can be measured in blood plasma, induced sputum and urine of asthmatic patients. PGE_2 is extracted from sputum using a specific immunoaffinity sorbent and concentrations are determined by enzyme immunoassay (EIA) (Aggarwal et al. 2010). The urinary metabolite, $9\alpha,11\beta$-prostaglandin F_2, can be measured as a marker of PGD_2 formation and mast cell activation (Misso et al. 2004, 2009, Aggarwal et al. 2010).

Measurement of Cysteinyl Leukotrienes

CysLT can be measured in blood plasma, induced sputum and urine of asthmatic patients. CysLT are extracted from sputum using a specific immunoaffinity sorbent and concentrations are determined by EIA (Aggarwal et al. 2010). The major urinary metabolite, LTE_4, can be measured as a marker of cysLT production in asthmatic patients (Misso et al. 2004, 2009, Aggarwal et al. 2010).

Anti-Prostaglandin Therapy

Cyclooxygenase inhibitors such as aspirin and non-steroidal anti-inflammatory drugs (NSAIDS) non-specifically block the formation of all prostaglandins and may therefore interfere with the beneficial anti-inflammatory effects of PGE_2. Furthermore, about 10% of asthmatic patients demonstrate the syndrome of aspirin-intolerant asthma following the use of aspirin and NSAIDS (Vally et al. 2002). Although these subjects tolerate selective COX-2 inhibitors such as celecoxib, inhibition of PGE_2 formation remains a concern (Daham et al. 2011). Therapeutic strategies that target the effects of specific prostaglandins will have to await the development of prostaglandin receptor antagonists and synthase inhibitors.

Anti-Leukotriene Therapy

Pharmacological agents that inhibit the production of leukotrienes or antagonize the function of the $CysLT_1$ receptor are in clinical use. Zileuton, an orally active inhibitor of 5-LO, inhibits the formation of LTB_4 and the cysLT. A number of orally active $CysLT_1$ receptor antagonists, the '-lukasts' are in clinical use as add-on or second-line therapy. However, clinical responsiveness to the $CysLT_1$ antagonists is heterogeneous (Scadding and Scadding 2010), with 34–53% of adults and up to 78% of children being non-responsive to montelukast. While anti-leukotrienes may be useful as add-on therapy in some patients with chronic asthma, there is insufficient evidence for their usefulness in acute asthma, and it is unclear whether they have a corticosteroid sparing effect (Scadding and Scadding 2010). A possible explanation for the variable efficacy of $CysLT_1$ receptor antagonists is that LTE_4 and its putative distinct receptors may have a more significant role in the pathophysiology of asthma than previously appreciated. The pharmacogenetic and environmental factors that might influence responsiveness to anti-leukotriene therapy require further investigation.

KEY FACTS ON PROSTAGLANDINS AND LEUKOTRIENES IN ASTHMA

- Research into prostaglandins and leukotrienes as mediators of inflammation in asthma dates back to the 1970s.
- Prostaglandins and leukotrienes are both derived from the 20 carbon polyunsaturated fatty acid, arachidonic acid, which is a component of cell membrane phospholipids.
- PGD_2 and PGE_2 are isomers, differing only in the position of the hydroxyl (C-5 in PGD_2, C-3 in PGE_2) and ketone (C-3 in PGD_2, C-5 in PGE_2) groups in the cyclopentane ring (Fig. 1).
- PGD_2 binds to two different receptor subtypes and PGE_2 binds to four different receptor subtypes.
- PGD_2 causes narrowing of the airways, whereas PGE_2 causes relaxation and opening of the airways.
- COX inhibitors such as aspirin and NSAIDS that non-specifically block the formation of all prostaglandins cause aspirin-intolerant asthma in about 10% of asthmatic patients.
- In cysteinyl leukotrienes the 20 carbon leukotriene lipid structure is conjugated to the amino acid, cysteine.
- Cysteinyl leukotrienes bind to at least two receptor subtypes, $CysLT_1$ and $CysLT_2$.
- Acting through the $CysLT_1$ receptor, LTD_4 causes maximum shortening of human airway smooth muscle at concentrations as low as 10^{-7} moles/L.
- $CysLT_1$ receptor antagonists are used as add-on therapy, but only about 50% of adult asthmatic patients show clinical responses to these drugs.

SUMMARY POINTS OF THE CHAPTER

- Prostaglandins and leukotrienes are key mediators of airway inflammation and remodelling in asthma.
- Cyclooxygenases together with prostaglandin synthases catalyse the biosynthesis of different prostaglandins, including PGD_2 and PGE_2, from arachidonic acid in cell membranes.
- Prostaglandins and leukotrienes exert their biological activities by binding to seven transmembrane, G protein-coupled receptors.

- PGD_2 binds to the DP and CRTH2 receptors to cause bronchoconstriction, vasodilatation, and the recruitment and activation of Th2 lymphocytes and eosinophils in the lungs of asthmatic patients.
- PGE_2 binds to EP (1–4) receptors, induces relaxation of airway smooth muscle, and has anti-inflammatory effects in the lungs of asthmatic patients.
- The biosynthesis of cysLT from arachidonic acid is catalysed by the enzymes, 5-lipoxygenase and LTC_4 synthase.
- CysLT bind to at least two receptors ($CysLT_1$ and $CysLT_2$), and binding to $CysLT_1$ results in contraction of airway smooth muscle and induction of airway hyperresponsiveness in asthmatic subjects.
- CysLT are also potent mediators of inflammation and airway remodelling in asthma.
- The balance between the production of cysLT and PGE_2 in the airways may be a determinant of chronic asthma severity, as well as airway inflammation and remodelling.
- $CysLT_1$ receptor antagonists are used in the treatment of asthma; however, clinical responsiveness to these drugs varies greatly.

DEFINITIONS AND EXPLANATION OF WORDS AND TERMS

Airway remodelling: Structural changes in the lungs of asthmatic patients that cause irreversible narrowing of the airways.

Antagonists: Molecules that block the binding of prostaglandins and leukotrienes to their specific receptors.

Arachidonic acid: A polyunsaturated fatty acid containing 20 carbon atoms and four double bonds that is the precursor of both prostaglandins and leukotrienes.

Bronchoconstriction: Contraction of airway smooth muscle, causing narrowing of the airways of asthmatic patients.

Cyclooxygenase: The enzyme catalyzing the formation of prostaglandins from arachidonic acid.

Eosinophil: A white blood cell that plays an important role in allergic asthma, and is a major source of cysLT.

G protein-coupled receptors: A family of transmembrane receptors that binds prostaglandins and leukotrienes on the cell surface and activates intracellular signalling pathways.

Induced sputum: Airway secretions coughed up by asthmatic patients after inhaling a vapour of 3 to 5% sodium chloride for 15 minutes.

Leukotriene C_4 synthase: The enzyme that catalyzes the formation of LTC_4, by conjugating the tripeptide, glutathione, to leukotriene A_4.

5-Lipoxygenase: The enzyme catalyzing the formation of leukotriene A_4 from arachidonic acid.

Macrophage: A white blood cell that engulfs and digests pathogens and cellular debris in the lungs and is also a major source of PGE_2.

Mast cell: An immune cell in the respiratory tract that releases LTC_4 and PGD_2 when activated by the binding of allergens to immunoglobulin E on the cell surface.

Th2 lymphocyte: A white blood cell that produces cytokines such as IL-4, IL-5 and IL-13, which play a major role in airway inflammation in asthmatic patients.

LIST OF ABBREVIATIONS

ATP	:	adenosine triphosphate
cAMP	:	cyclic adenosine monophosphate
COX	:	cyclooxygenase
cPGES	:	cytosolic prostaglandin E synthase
cysLT	:	cysteinyl leukotriene
$CysLT_1$	:	cysteinyl leukotriene receptor 1
$CysLT_2$	:	cysteinyl leukotriene receptor 2
DAG	:	diacylglycerol
EIA	:	enzyme immunoassay
FLAP	:	5-lipoxygenase activating protein
GTP	:	guanosine triphosphate
IL	:	interleukin
IP_3	:	inositol 1,4,5-trisphosphate
5-LO	:	5-lipoxygenase
LT	:	leukotriene
mPGES	:	membrane-bound prostaglandin E synthase
NSAID	:	non-steroidal anti-inflammatory drug

PG	:	prostaglandin
Th2	:	T helper type 2
Tx	:	thromboxane

REFERENCES

Aggarwal, S., Y.P. Moodley, P.J. Thompson and N.L. Misso. 2010. Prostaglandin E2 and cysteinyl leukotriene concentrations in sputum: association with asthma severity and eosinophilic inflammation. Clin Exp Allergy 40: 85–93.

Austen, K.F., A. Maekawa, Y. Kanaoka and J.A. Boyce. 2009. The leukotriene E4 puzzle: finding the missing pieces and revealing the pathobiologic implications. J Allergy Clin Immunol 124: 406–414.

Barnes, P.J., K.F. Chung and C.P. Page. 1998. Inflammatory mediators of asthma: an update. Pharmacol Rev 50: 515–596.

Daham, K., W.L. Song, J.A. Lawson, M. Kupczyk, A. Gülich, S.E. Dahlén, G.A. FitzGerald and B. Dahlén. 2011. Effects of celecoxib on major prostaglandins in asthma. Clin Exp Allergy 41: 36–45.

Fujii, M., H. Tanaka and S. Abe. 2005. Interferon-γ upregulates expression of cysteinyl leukotriene type 2 receptors on eosinophils in asthmatic patients. Chest 128: 3148–3155.

Gervais, F.G., R.P. Cruz, A. Chateauneuf, S. Gale, N. Sawyer, F. Nantel, K.M. Metters and G.P. O'Neill. 2001. Selective modulation of chemokinesis, degranulation, and apoptosis in eosinophils through the PGD2 receptors CRTH2 and DP. J Allergy Clin Immunol 108: 982–988.

Hamid, Q. and M. Tulic. 2009. Immunobiology of asthma. Annu Rev Physiol 71: 489–507.

Hata, A.N. and R.M. Breyer. 2004. Pharmacology and signaling of prostaglandin receptors: Multiple roles in inflammation and immune modulation. Pharmacol Ther 103: 147–166.

Holgate, S.T., M. Peters-Golden, R.A. Panettieri and W.R. Henderson, Jr. 2003. Roles of cysteinyl leukotrienes in airway inflammation, smooth muscle function, and remodeling. J Allergy Clin Immunol 111: S18–S36.

Kabashima, K. and S. Narumiya. 2003. The DP receptor, allergic inflammation and asthma. Prostaglandins Leukot Essent Fatty Acids 69: 187–194.

Lee, T.H., G. Woszczek and S.P. Farooque. 2009. Leukotriene E4: perspective on the forgotten mediator. J Allergy Clin Immunol 124: 417–421.

Knight, D.A., G.A. Stewart and P.J. Thompson. 1995. Prostaglandin E2, but not prostacyclin inhibits histamine-induced contraction of human bronchial smooth muscle. Eur J Pharmacol 272: 13–19.

Matsuoka, T., M. Hirata, H. Tanaka, Y.Takahashi, T. Murata, K. Kabashima, Y. Sugimoto, T. Kobayashi, F. Ushikubi, Y. Aze, N. Eguchi, Y. Urade, N. Yoshida, K. Kimura, A. Mizoguchi, Y. Honda, H. Nagai and S. Narumiya. 2000. Prostaglandin D2 as a mediator of allergic asthma. Science 287: 2013–2017.

Misso, N.L., S. Aggarwal, S. Phelps, R. Beard and P.J. Thompson. 2004. Urinary leukotriene E4 and 9α,11β-prostaglandin F2 concentrations in mild, moderate and severe asthma, and in healthy subjects. Clin Exp Allergy 34: 624–631.

Misso, N.L., S. Aggarwal, P.J. Thompson and H. Vally. 2009. Increases in urinary 9α,11β-prostaglandin F_2 indicate mast cell activation in wine-induced asthma. Int Arch Allergy Immunol 149: 127–132.

Newcomer, M.E. and N.C. Gilbert. 2010. Location, location, location: compartmentalization of early events in leukotriene biosynthesis. J Biol Chem 285: 25109–25114.

Oguma, T., L.J. Palmer, E. Birben, L.A. Sonna, K. Asano and C.M. Lilly. 2004. Role of prostanoid DP receptor variants in susceptibility to asthma. N Engl J Med 351: 1752–1763.

Parachuri, S., H. Tashimo, C. Feng, A. Maekawa, W. Xing, Y. Jiang, Y. Kanaoka, P. Conley and J.A. Boyce. 2009. Leukotriene E4-induced pulmonary inflammation is mediated by the P2Y12 receptor. J Exp Med 206: 2543–2555.

Peacock, C.D., N.L. Misso, D.N. Watkins and P.J. Thompson. 1999. PGE2 and dibutyryl cyclic adenosine monophosphate prolong eosinophil survival *in vitro*. J Allergy Clin Immunol 104: 153–162.

Peters-Golden, M. and W.R. Henderson Jr. 2007. Leukotrienes. N Engl J Med 357: 1841–1854.

Petrovic, N., D.A. Knight, J.S. Bomalaski, P.J. Thompson and N.L. Misso. 2006. Concomitant activation of extracellular signal-regulated kinase and induction of COX-2 stimulates maximum prostaglandin E2 synthesis in human airway epithelial cells. Prostaglandins Other Lipid Mediat 81: 126–135.

Pettipher, R., T.T. Hansel and R. Armer. 2007. Antagonism of the prostaglandin D2 receptors DP1 and CRTH2 as an approach to treat allergic diseases. Nat Rev Drug Discov 6: 313–325.

Samuelsson, B., R. Morgenstern and P-J. Jakobsson. 2007. Membrane prostaglandin E synthase-1: a novel therapeutic target. Pharmacol Rev 59: 207–224.

Scadding, G.W. and G.K. Scadding. 2010. Recent advances in antileukotriene therapy. Curr Opin Allergy Clin Immunol 10: 370–376.

Singh, R.K., S. Gupta, S. Dastidar and A. Ray. 2010. Cysteinyl leukotrienes and their receptors: molecular and functional characteristics. Pharmacology 85: 336–349.

Tilley, S.L., T.M. Coffman and B.H. Koller. 2001. Mixed messages: modulation of inflammation and immune responses by prostaglandins and thromboxanes. J Clin Invest 108: 15–23.

Urade, Y. and N. Eguchi. 2002. Lipocalin-type and hematopoietic prostaglandin D synthases as a novel example of functional convergence. Prostaglandins Other Lipid Mediat 68–69: 375–382.

Vally, H., M.L. Taylor and P.J. Thompson. 2002. The prevalence of aspirin intolerant asthma (AIA) in Australian asthmatic patients. Thorax 57: 569–574.

Vancheri, C., C. Mastruzzo, M.A. Sortino and N. Crimi. 2004. The lung as a privileged site for the beneficial actions of PGE_2. Trends Immunol 25: 40–46.

Zhu, J., Y. Qiu, D. Figueroa, V. Bandi, H. Galczenski, K. Hamada, K. Guntupalli, J. Evans and P. Jeffery. 2005. Localization and upregulation of cysteinyl leukotriene-1 receptor in asthmatic bronchial mucosa. Am J Respir Cell Mol Biol 33: 531–540.

3

Unsatisfactory Asthma Control: A Review

Braido Fulvio,[1,a,*] Baiardini Ilaria,[1,b] Lagasio Chiara,[1,c] Sclifò Francesca,[1,d] Lan-Anh Le[2] and Canonica Giorgio Walter[1,e]

ABSTRACT

Despite clinical research showing that good asthma control is an achievable target, real life studies suggest that a significant proportion of patients still suffer from symptoms and lifestyle restrictions, with a considerable impact on quality of life and associated societal cost. Achieving good asthma control is dependent on a number of interacting variables including disease severity and both patients' and physicians' knowledge and behaviour. The later may be impaired by insufficient use of validated tools for assessment of symptom control in daily clinical practice, and may be relevant in understanding why management goals are often not achieved.

[1]University of Genoa, Allergy and Respiratory Diseases, Department of Internal Medicine, Pad. Maragliano, Largo Rosanna Benzi 10, 16132 Genova, Italy.
[a]Email: fulvio.braido@unige.it
[b]Email: ilaria.baiardini@libero.it
[c]Email: chiaralagasio@libero.it
[d]Email: fra.scli@libero.it
[e]Email: canonica@unige.it
[2]Rosemead Surgery, 8a Ray Park Avenue, Maidenhead, Berkshire, SL6 8DS, UK; Email: organisedlan@gmail.com
*Corresponding author

List of abbreviations after the text.

INTRODUCTION

For asthma, as with most chronic diseases, the goal of treatment is to control the disease process itself. There are many classification systems for asthma control, however currently no one system has been universally adopted. Asthma control can be split into two distinct categories (Reddel et al. 2009): current clinical control (including the avoidance of asthma symptoms, minimizing reliever use, allowing normal levels of activity and achieving normal lung function tests) and future prevention (which includes prevention of asthma exacerbations and the maintenance of normal lung function tests in the absence of drug side-effects).

The Gaining Optimal Asthma ControL study (GOAL) (Bateman et al. 2004), defined asthma control based on the treatment goals of the Global Initiative for Asthma/National Institutes of Health guidelines (GINA) (Global Initiative for Asthma 1997). Patients were considered "totally-controlled" if, during the 8 consecutive assessment weeks, they had no exacerbations, no emergency room attendances, and no medication related adverse events. "Well-controlled asthma" was achieved if in 7 out of the 8 weeks patients were asymptomatic for at least 5 days per week, and they only required rescue medication during 2 days (or on 4 separate occasions) each week. If any asthma exacerbations, emergency room visits, or medication adverse events occurred during this 8 week period the patient's asthma control would fail to meet the criteria for control, irrespective of how well their asthma was controlled the rest of the time during the 8 weeks. Beyond revealing the greater efficacy of the combined use of inhaled corticosteroid (ICS) plus long acting β-2 agonist (LABA) compared to ICS alone, this study provided important knowledge regarding asthma management. The goal of total control was achievable in a fair percentage of patients, regardless of disease severity, and increased with treatment time. A post-hoc analysis (O'Byrne et al. 2008) of the Formoterol And Corticosteroid Establishing Therapy (FACET) trial confirmed that in patients with moderate-to-severe asthma, sustained guideline-defined asthma control was possible in a high proportion of patients. These studies all suggest that with the use of the currently available asthma drugs, good control of asthma is possible in most patients. However, these results obtained in clinical trials where highly motivated and selected patients are carefully followed by a team of researchers, still need to be translated into real-life practice.

THE ASSESSMENT OF ASTHMA CONTROL: VALIDATED TOOLS

Since the focus of asthma management has shifted from treating acute attacks to achieving asthma control, the need for tools that can reliably identify the level of control itself has become increasingly important. Single clinical and functional parameters seem insufficient to ensure a proper asthma control assessment. For instance, an observational study demonstrated that measuring the changes in asthma control only by spirometry, morning and evening Peak Expiratory Flows (PEFs) and clinical judgement may be inappropriate, because of an unexpected tendency by physicians to overestimate improvement and to underestimate deteriorations in asthma control (Juniper et al. 2004).

Validated scales to quantify asthma control are now available, including the Asthma Control Questionnaire (ACQ) (Juniper et al. 2005), the Asthma Control Test (ACT) (Nathan et al. 2001), the Asthma Therapy Assessment Questionnaire (ATAQ) (Skinner et al. 2004) and the Perceived Control of Asthma Questionnaire (PCAQ) (Katz et al. 2002).

The ACQ (Juniper et al. 2005) originally contained seven items including the highest score across 5 symptoms, predicted FEV_1% and daily rescue bronchodilator use. However, three shortened versions (symptoms alone, symptoms plus FEV1 and symptoms plus short-acting β2 agonists use (SABA)) have since also been validated. All four versions of ACQ have demonstrated reliability, validity, and sensitivity to change, with even the smallest change in score of 0.5 considered to be a clinically significant improvement or deterioration in disease control.

The Asthma Control Test (ACT) (Nathan et al. 2004) is a patient-completed questionnaire of five items with 5 response options investigating limitations at work or school due to asthma, the presence of daytime or nighttime symptoms, the use of rescue medications and the subjective perception of the level of asthma control in the previous four weeks. The sum of the scores allows asthma control to be categorized as follows: non-controlled asthma (5–19 points); controlled asthma (20–24 points) and optimal disease control (25 points).

The Asthma Therapy Assessment Questionnaire (ATAQ) (Skinner et al. 2006) is a 20-item parent-completed questionnaire for paediatric populations of 5–17 years, that generates indicators of potential problems in different categories including symptom control, behaviour and attitude barriers, self-efficacy barriers, and communication issues.

The Perceived Control of Asthma Questionnaire (PCAQ) is an 11-item tool that assesses the patients' perceptions of their ability to manage asthma and its exacerbations. Responses are graded on a 5-point scale, scoring between 11 and 55, with higher scores reflecting greater perceived control of asthma (Katz et al. 2002).

A retrospective analysis (O'Byrne et al. 2010) compared assessment of asthma control using the ACQ, GINA or GOAL study criteria in a large population. The GINA and GOAL criteria identified a similar proportion of patients within each asthma control classification. The GINA and GOAL control strata were similar in terms of ACQ scores. This analysis showed that the percentages of patients considered by GINA criteria to have controlled and partly controlled asthma and by GOAL criteria to have totally controlled and well-controlled asthma were comparable to an ACQ score of 1.00.

Similar results were found comparing GOAL criteria with the ACQ (Juniper et al. 2006). For all the ACQ versions, the crossover point between well-controlled and not well-controlled is close to 1.00. More precisely, if a patient has an ACQ score of 0.75 or less, there is an 85% chance that his asthma is well-controlled and if a patient has an ACQ score of 1.50 or greater, there is an 88% chance that his asthma is not well-controlled.

A study of the relationship between the GINA criteria and the ACT (Thomas et al. 2009) showed that an ACT score lower than 19 was able to predict uncontrolled asthma as defined by GINA. Another study assessed three different guideline-based tools (GINA, the National Asthma Education and Prevention Program and the Joint Task Force Practice Parameter) against the ACQ and ACT. Agreement was high among the guidelines tools, moderate between the ACQ and the ACT, and poor to fair between the ACT or the ACQ and the other 3 tools (Khalili et al. 2008).

ASTHMA CONTROL IN REAL-LIFE

Large population studies have clearly demonstrated that asthma control is not always achieved. Asthma Insights and Reality in Europe (AIRE) study (Rabe et al. 2000), involving over 2,800 European patients, showed that many patients were far from asymptomatic: 46% of patients reported daytime symptoms and 30% asthma-related sleep disturbances at least once a week. Within the last year of the study, 25% underwent unscheduled urgent care visits, 10% had one or more emergency room visits and 7% had overnight hospitalization due to asthma. Similarly in Asia (Lai et al. 2003) asthma control is far from optimal. In a population of 2,323 adults and 884 children, 51.4% of patients reported daytime symptoms, 44.3% had night awakenings due to asthma symptoms during the last four weeks and 44.7% patient's experienced limitations in physical activity. Moreover 26.5% of adults and 36.5% of children reported work or school absence in the past year. A high percentage of respondents (43.6%) had been hospitalised or had made emergency, unplanned visits due to acute asthma exacerbations.

Furthermore, a cross-sectional study performed in 29 countries on 7,786 adults and 3,153 children showed that asthma control worldwide is far from the goals suggested by GINA, with a significant proportion of patients having symptoms, lifestyle restrictions and requiring emergency care. The use of anti-inflammatory preventative medication, even in patients with severe persistent asthma was low, ranging from 26% in Western Europe to 9% in Japan (Rabe et al. 2004).

The European Community Respiratory Health Survey II showed that only 15% of subjects who had used ICSs in the last year and 45% of non-ICS users had their asthma under control (Cazzoletti et al. 2007).

In the INSPIRE (INternational aSthma Patient Insight REsearch) study 3,415 treated asthma adults were interviewed by phone (Partridge et al. 2006). Despite being prescribed regular maintenance therapy, 74% of patients used SABA daily and 51% were classified by the ACQ as having uncontrolled asthma. The European National Health and Wellness Survey showed that a substantial proportion of patients with uncontrolled asthma (50.4%) were older, had lower education levels, and were more likely to be obese, depressed and smoke, when compared to those considered to have controlled asthma (Demoly et al. 2009).

THE RELATIONSHIP BETWEEN CONTROL AND HEALTH-RELATED QUALITY OF LIFE

It is reasonable to hypothesize an association between better asthma control and a reduced burden on health-related quality of life (HRQoL), but few studies have specifically addressed this relationship using validated tools.

Clinical Trials

A retrospective meta-analysis (Bateman et al. 2004) of three randomized double blind placebo controlled studies, aimed to evaluate the effect of salmeterol/fluticasone propionate combination therapy on HRQoL of patients with different levels of asthma severity. The results showed that the patients who achieved disease control according to GINA guidelines reported better HRQoL than patients who did not achieve control and indeed these levels were similar to healthy control subjects. Moreover, HRQoL correlated not only with the level of control, but also with the treatment allocation: patients treated with salmeterol/fluticasone propionate combination therapy reported a significant improvement in HRQoL scores compared to other treatment options, irrespective of the level of control. This suggests that it is possible to reduce the impact of asthma on the patient's life with appropriate drug therapy, even if optimal control is not achieved.

In particular, the correlation between asthma control and HRQoL has been analysed in a cohort of more than 3000 patients involved in the GOAL (Bateman et al. 2007) study. The mean HRQoL scores at end-point were significantly higher in patients achieving total control compared with those achieving good control, and also higher in those achieving good control compared to patients who did not achieve control.

The number of patients reporting a clinically significant HRQoL improvement was higher in the group treated with the combination treatment than ICS alone. Moreover, in about half of the patients, when treatment was aimed at achieving total control, HRQoL scores were almost maximal, with the majority of patients having minimal disturbances from their asthma and higher levels of well being.

Real-life Studies and Surveys

The association between asthma control and HRQoL has also been demonstrated in real-life studies. In a study (Braido et al. 2009) involving patients with both asthma and rhinitis, only 56% of subjects reached an ACT score ≥ 20. Patients with controlled asthma showed significantly better HRQoL compared with uncontrolled patients. However, despite HRQoL in asthma improving when patients were controlled, optimal scores were not always seen, i.e., the achievement of asthma control did not necessarily equate to the achievement of maximal HRQoL (Fig. 1).

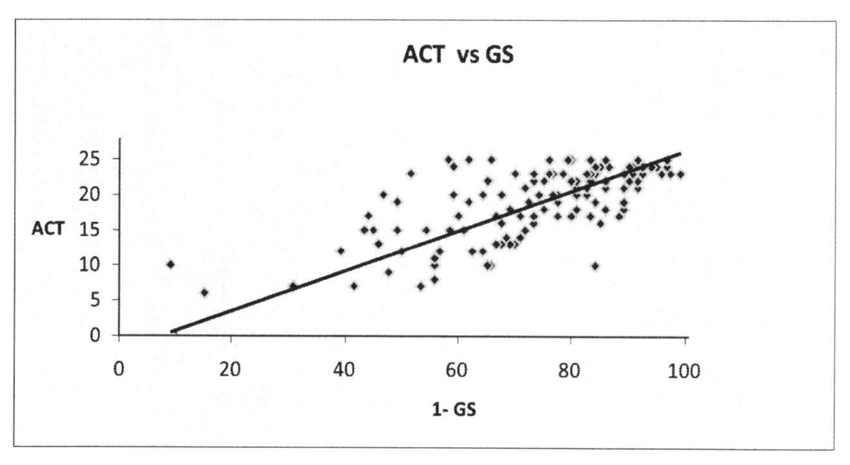

Figure 1 Relationship between asthma control and quality of life.
ACT: Asthma Control Test
GS: Global Summary of Rhinasthma questionnaire

The role of rhinitis in determining HRQoL has also been explored. Irrespective of ACT scores, and even in patients with controlled asthma, those with additional rhinitis symptoms had a worse HRQoL.

A significant difference between patients with controlled and uncontrolled asthma has been further demonstrated in health status, with worse physical and mental summary scores from the generic SF-12 survey in patients with ACT score ≤ 19 (Demoly et al. 2010). The results of these studies have been confirmed in large cohorts including paediatric populations. Recent surveys (Guilbert et al.

2010, Dean et al. 2009, Dean et al. 2010) showed that also children with uncontrolled asthma had significantly lower HRQoL than those who achieved control.

In order to explore the relationship between the perceived asthma control (assessed by PCAQ) and patient's health outcomes, Calfee et al. (2006) performed a study on 865 patients. Structured telephone interviews were conducted for patients in a prospective cohort who were observed for hospitalisations due to asthma. Greater perceived control resulted in improved health status and HRQoL.

REASONS FOR POOR CONTROL

The achievement of asthma control can be considered as the result of the interaction among different variables such as the disease, the patient themselves and the doctor's input (Baiardini et al. 2009) (Table 1). Some of these factors will be detailed in the following paragraphs. It is important to remember that some factors linked to the guidelines themselves (**i.e., complexity**), to their implementation (**communication strategies**) and to the socio-cultural context (**i.e., system efficiency**) represent barriers that limit the achievement of the guidelines' goals and, therefore, the improvement of asthma control.

An Example of Disease-related Factors that Influence Asthma Control: the Role of Rhinitis

Allergic rhinitis is not only a burden to the patient and their health care system, but it represents an additional asthma-related cost. Its role must be taken into account in asthma management especially with regard to treatment and prophylaxis. This concept has been clearly demonstrated by Allergic Rhinitis and Its Impact on Asthma (ARIA) (Bousquet et al. 2001), and in further clinical trials. Among patients suffering from asthma and concomitant allergic rhinitis, the risk of asthmatic crisis requiring emergency medical assistance or hospitalization, was significantly lower in subjects treated for allergic rhinitis compared to those whose allergic rhinitis was not treated (Ponte et al. 2008). These data support the hypothesis that optimal

Table 1 Factors that influence the achievement of asthma control.

Reasons of poor control	Variables	Examples
Disease-related	Comorbidities	Rhinitis, rhinosinusitis, gastro-oesophageal reflux, obstructive sleep apnoea
	Exacerbating factors	Ongoing occupational or allergen exposures
Patient-related	Socio-demographic factors	Female sex, education below secondary level, elderly age
	Adherence	Undertreatment, overtreatment, irregular visits to healthcare providers, insufficient monitoring of symptoms, no lifestyle modifications
	Psychiatric comorbidity	Anxiety, depressive disorders
	Psychological characteristics	Alexithymia (a personality trait characterized by difficulty in identifying and verbally expressing feeling)
	Perceptions	Tendency to tolerate symptoms, exacerbations and lifestyle limits as an inevitable consequence of asthma and show
	Expectations	Low expectations about the degree of control that is possible
	Behaviours	Smoking habits Poor inhaler technique
	Knowledge	Lack of information about the disease and its treatment
Doctor-related	Underdiagnosis	Limited awareness of asthma prevalence
	Knowledge of current guidelines	Lack of consciousness and familiarity about guidelines availability
	Attitude towards guidelines	Difficulty in accepting a particular document, or the concept itself of the guidelines Lack of confidence in personal abilities to put the recommendations into practice Expectations of failure in following guidelines
	Guidelines implementations	Difficulty changing deep-seated routines

rhinitis management can lead to the improvement in concomitant asthma and, therefore, a combined treatment strategy should be planned in order to achieve the best possible health status.

Examples of Patient-related Factors that Influence Asthma Control: Adherence and Knowledge

Despite the availability of effective diagnostic tools and safe therapeutic drugs which should allow the satisfactory identification, monitoring and management of asthma, these alone are not sufficient to obtain disease control. It is the active involvement of patients in their disease and their therapy and management that is essential in reaching and maintaining asthma control.

As with other chronic diseases, a high percentage of asthmatic patients incorrectly follow their doctor's prescriptions, resulting in underdosing, overdosing, or simply incorrect drug usage. Although poor control is the commonest consequence, others include an increase in side effects, disease progression, more complications, the need for unplanned visits, the requirement of other diagnostic investigations and the unnecessary use of stronger drugs. The numerous causes of non-adherence are generally grouped into four larger categories (Table 2). Since many factors influence adherence, interventions focused solely on one adherence factor struggle to demonstrate significant benefits.

Some factors when altered such as patient's beliefs, knowledge and expectations, and the doctor-patient relationship can result in the improvement of clinical outcomes. In particular, patients on long term treatment were sometimes found to have inadequate knowledge about their asthma. A survey (Baiardini et al. 2006) exploring the reasons for non-adherence showed that 25% of patients didn't know the symptoms of asthma; 25% were not taking their medication correctly, 19% were not able to identify worsening signs, and only 40% of patients studied were able to monitor clinical parameters. Although the majority of patients appreciated the necessity for drugs, at least 28% of them were worried about side effects.

Non modifiable variables include difficulties linked to a patient's age, cognitive deficits, disease characteristics and the necessity for long term treatment.

Table 2 Causes of non-adherence.

Factors linked to the patient	• presence of physical disorders • cognitive difficulties • psychiatric comorbidities • age (children, adolescents and elderly present high risk of non-adherence) • knowledge • expectations • social and family support • coping style
Variables linked to the disease	• chronicity • symptom stability • absence of symptoms
Variables linked to the treatment	• high number of daily doses • presence of side effects • complexity of the therapeutic regimes • ease of use • costs
Variables related to the doctor-patient relationship	• bad relationship • inappropriate doctor or patient behaviour

It is of great importance that doctors do not underestimate the influence of adherence on achievement of asthma control and other clinical outcomes.

Examples of Doctor-related Factors that Influence Asthma Control: Knowledge, Strategies and Behaviours

Two recent questionnaire based studies offered glimpses of links between physicians' behaviour and the doctor patient relationship and how this affects patients' knowledge of their own disease, suggesting important elements that may influence asthma management.

In the first study (Braido et al. 2010), 811 general practitioners (GPs) and 230 specialists in respiratory medicine, who attended a continuing medical education program, completed a questionnaire on aspects related to the pathogenesis of asthma and its control, the applicability of research and guidelines in daily practice, and the doctor-patient relationship. Barriers to achieving asthma control (Fig. 2) included the limited level of knowledge among GPs and specialists, regarding the use of instruments such as the ACT which

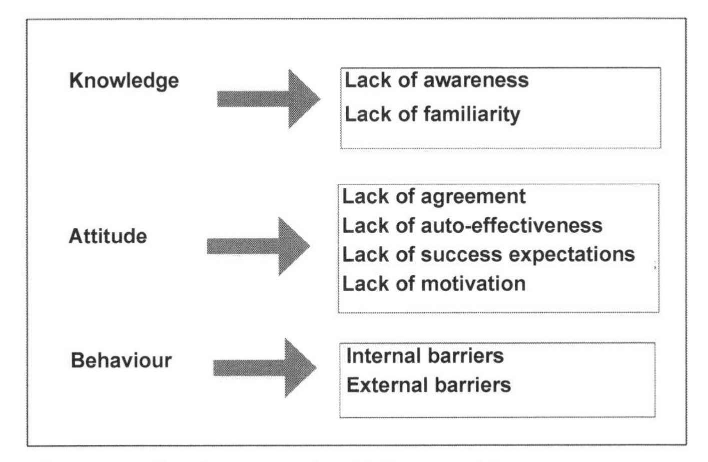

Figure 2 Physician-related reasons of guidelines avoidance.

was only used by 20% of GPs and 43% of specialists. When asked to identify the level of asthma control, only 20% provided correct answers. Although over 90% of physicians considered chronic inflammation to be the main characteristic of asthma, up to 40% of patients were considered to not require long-term treatment. Both GPs and specialists preferred fixed dose continuous regimes (58% and 54%, respectively) and believed that a self-management plan was feasible only in a very small percentage of patients. Another noteworthy finding was that neither GPs nor specialists had complete trust in the applicability of clinical trial findings and guidelines in real life.

Regarding the doctor-patient relationship, less than one third of GPs and only a fifth of specialists adopted a cooperative approach (aimed at actively involving the patient in the entire treatment process and building a partnership), being instead, and particularly in the case of GPs, "paternalistic". This approach is not ideal for management of a chronic disease such as asthma, since it involves a passive role on the part of the patient with limited autonomy. For both GPs and specialists, the level of patient education was the key determinant of adherence. Patient expectations and treatment peculiarities were considered less important aspects. The findings of this survey provide, on the one hand, an explanation for the poor level of asthma control seen in daily practice, and on the other, evidence for the further need for the development of targeted information and patient education programs.

In the second survey (Braido et al. 2011), GPs involved in asthma medical education programs were asked to answer questions concerning their experience on clinical aspects of asthma management (pathogenesis and control, applicability of research and guidelines to daily practice) and their doctor patient relationships and patient educational strategies. Each physician recruited 3 asthma patients who each indicated 3 aspects of their disease (from a choice of 10 possible options) that they felt they were less informed about. A total of 2,332 GPs and 7,884 patients participated in the study. Only a half of GPs thought that control could be reached independently of disease severity, and a relevant percentage of physicians minimized their responsibility in unsatisfactory asthma control, declaring that control could be reached only by more scrupulous patients. Most GPs thought that a long term treatment was applicable in the majority of patients but 1 out 4 believed that this strategy was applicable only in patients with specific characteristics.

Surprisingly, the topic patients chose as feeling least informed about was "the meaning of asthma control" and this choice seems to be significantly associated to physician's behaviour.

These patients were less aware of the meaning of asthma control, when their physicians did not explore how the patient perceived their own disease, failed to provide additional supporting educational material, did not listen to the patient and did not involve the patient in their ongoing asthma management plan. It is relevant to notice that this topic was ranked lower at 8th place, when including specialists and patient's belonging to patients' associations. This demonstrates that physicians tend to take for granted some important aspects of patient education in asthma management. If patients are not sufficiently educated and do not understand the meaning of control, this facet of doctor patient interaction may be a further barrier to their education and by extension their optimal management.

PRACTICE AND PROCEDURES

Asthma control should be assessed using validated tools taking into account the following considerations:

1) The practical application of the tool should comply with the developers' intentions and instructions, i.e., the patient and not

the physician or caregiver should complete the form and the questionnaire should not be conducted over the telephone or mailed (unless validated for these methods). It should only be used for the patient population it was designed and validated for, e.g., tools for adults are inappropriate for adolescents.

2) The questionnaire, including instructions for use and response methods, should not be modified or adapted in any way.
3) Potential copyright regulations or commercial fees related to the tool should be examined before using a questionnaire.
4) When using a tool in a language that differs from the original, it should be taken into account that a simple translation is not sufficient, since a back-translation and a cross-cultural validation must be performed.

KEY FACTS ABOUT ASTHMA

- Asthma is a chronic inflammatory disorder that impacts patient's quality of life. The aims of the treatment are to reduce symptoms and disease burden.
- From a patients' perspective, the impact of asthma symptoms on quality of life is the most important aspect of the disease and its treatment.
- In order to accurately assess the impact of asthma on patients, it is advisable to consider composite scoring rather than a single feature or parameter assessment.
- Disease control remains core in defining asthma management by international guidelines.
- The severity of the disease requires proportional levels of drug treatment in order to achieve control.

SUMMARY POINTS

- In real life settings, asthma control is not always achieved.
- Several real life surveys showed that at least 50% patients reported daytime symptoms.
- Proper diagnosis and patient follow up are needed in order to plan an effective treatment schedule.

- Uncontrolled asthma occurs despite the patients being prescribed regular maintenance therapy.
- Patients with uncontrolled asthma compared to those with controlled asthma tend to be older, obese, depressed and smoke with a lower education level.
- Large clinical trials show an association between asthma control and HRQoL.
- HRQoL is higher in patients achieving total control compared with those achieving good control, and in patients achieving good control compared to patients who are not well controlled.
- HRQoL of controlled asthma patients was very similar to healthy control subjects.
- When treatment is aimed at achieving total control, the majority of patients have the possibility to reach high HRQoL levels, minimizing the influence of asthma on their life.
- Both adults and children with controlled asthma have significantly better HRQoL scores compared with uncontrolled patients.
- In real life, irrespective of their ACT score, patients with rhinitis symptoms showed worse HRQoL. Furthermore, in patients with controlled asthma, the presence of rhinitis symptoms was associated with a poorer HRQoL.
- Unsatisfactory asthma control may be considered as the result of the interaction among different variables.
- Disease-related reasons for unsatisfactory control include the presence of comorbidity and the presence of exacerbating factors.
- Patient-related reasons for unsatisfactory control include psychological and socio-demographic characteristics, non-adherence to treatment, the patient's expectations, perceptions, behaviours and knowledge.
- Doctor-related reasons for unsatisfactory control include under-recognition of symptoms, inadequate assessment of asthma control, and difficulty in following guidelines.

DEFINITION AND EXPLANATION OF WORDS AND TERMS

Adherence: The rate of concordance between physicians' suggestions and patients' application to a treatment plan.

Asthma: Chronic inflammatory disorder characterized by episodes of airways obstruction.

Clinical trial: study to identify the efficacy or safety of a diagnostic or treatment intervention.

Control tools: Validated instruments aimed to quantify disease control.

Control: The total avoidance of disease impact on patients' life.

Direct costs: Money expenditure directly related to a disease (i.e., drugs).

Indirect costs: Money expenditure indirectly related to a disease (i.e., lost working days).

Quality of Life: The impact of the disease and its treatment from the patients' perspective.

Real-life survey: Observational study aimed to evaluate a phenomenon in real-life.

Severity: The degree of symptoms and disease features.

LIST OF ABBREVIATIONS

ARIA	:	Allergic Rhinitis and Its Impact on Asthma
ACQ	:	Asthma Control Questionnaire
ACT	:	Asthma Control Test
AIRE	:	Asthma Insights and Reality in Europe
ATAQ	:	Asthma Therapy Assessment Questionnaire
FEV_1	:	Forced Expiratory Volume in one second
FACET	:	Formoterol and Corticosteroid Establishing Therapy
GOAL	:	Gaining Optimal Asthma ControL
GINA	:	Global Initiative for Asthma
GP	:	General Practitioner

HRQoL	:	Health Related Quality of Life
ICS	:	Inhaled Corticosteroid
INSPIRE	:	INternational aSthma Patient Insight REsearch
LABA	:	Long Acting β2 Agonist
PEF	:	Peak Expiratory Flow
PCAQ	:	Perceived Control of Asthma Questionnaire
SABA	:	Short Acting β2 Agonist

REFERENCES

Baiardini, I., F. Braido, A. Giardini, G. Majani, C. Cacciola, A. Rogaku, A. Scordamaglia and G.W. Canonica. 2006. Adherence to treatment: assessment of an unmet need in asthma. J Investig Allergol Clin Immunol 16: 218–23.

Baiardini, I., F. Braido, M. Bonini, E. Compalati and G.W. Canonica. 2009. Why do doctors and patients not follow guidelines? Curr Opin Allergy Clin Immunol 9: 228–33.

Bateman, E.D., H.A. Boushey, J. Bousquet, W.W. Busse, T.J. Clark, R.A. Pauwels and S.E. Pedersen; GOAL Investigators Group. 2004. Can guideline-defined asthma control be achieved? The Gaining Optimal Asthma ControL study. Am J Respir Crit Care Med 170: 836–44.

Bateman, E.D., J. Bousquet, M.L. Keech, W.W. Busse, T.J. Clark and S.E. Pedersen. 2007. The correlation between asthma control and health status: the GOAL study. Eur Respir J 29: 56–62.

Bousquet, J., P.B. van Cauwenberge, N. Khaltaev, ARIA Workshop Group and WHO. 2001. Allergic rhinitis and its impact on asthma. J Allergy Clin Immunol 108: S1–S334.

Braido, F., I. Baiardini, E. Stagi, M.G. Piroddi, S. Balestracci and G.W. Canonica. 2010. Unsatisfactory asthma control: astonishing evidence from general practitioners and respiratory medicine specialists. J Investig Allergol Clin Immunol 20: 9–12.

Braido, F., I. Baiardini, S. Balestracci, V. Ghiglione, E. Stagi, E. Ridolo, R. Nathan and G.W. Canonica. 2009. Does asthma control correlate with quality of life related to upper and lower airways? A real life study. Allergy 64: 937–43.

Braido, F., I. Baiardini, S. Menoni, V. Brusasco, S. Centanni, G. Girbino, R. Dal Negro and G.W. Canonica. 2011. Asthma management failure: a flaw in physician's behavior or in patient's knowledge? J Asthma 48: 266–74.

Calfee, C.S., P.P. Katz, E.H. Yelin, C. Iribarren and M.D. Eisner. 2006. The influence of perceived control of asthma on health outcomes. Chest 130: 1312–8.

Cazzoletti, L., A. Marcon, C. Janson, A. Corsico, D. Jarvis, I. Pin, S. Accordini, E. Almar, M. Bugiani, A. Carolei, I. Cerveri, E. Duran-Tauleria, D. Gislason, A. Gulsvik, R. Jõgi, A. Marinoni, J. Martínez-Moratalla, P. Vermeire and R. de Marco; Therapy and Health Economics Group of the European Community Respiratory Health Survey. 2007. Asthma control in Europe: a real-world evaluation based on an international population-based study. J Allergy Clin Immunol 120: 1360–7.

Dean, B.B., B.C. Calimlim, P. Sacco, D. Aguilar, R. Maykut and D. Tinkelman. 2010. Uncontrolled asthma: assessing quality of life and productivity of children and their caregivers using a cross-sectional Internet-based survey. Health Qual Life Outcomes. 8: 96.

Dean, B.B., B.M. Calimlim, S.L. Kindermann, R.K. Khandker and D. Tinkelman. 2009. The impact of uncontrolled asthma on absenteeism and health-related quality of life. J Asthma 46: 861–6.

Demoly, P., B. Gueron, K. Annunziata, L. Adamek and R.D. Walters. 2010. Update on asthma control in five European countries: results of a 2008 survey. Eur Respir Rev 19: 150–7.

Demoly, P., P. Paggiaro, V. Plaza, S.C. Bolge, H. Kannan, B. Sohier and L. Adamek. 2009. Prevalence of asthma control among adults in France, Germany, Italy, Spain and the UK. Eur Respir Rev 18: 105–12.

[GINA] Global Initiative for Asthma. 1997. National Asthma Education and Prevention Program. Guidelines for the diagnosis and management of asthma: expert panel report 2. Bethesda: National Institutes of Health, National Heart, Lung and Blood Institute; Publication No. 97–4051.

Guilbert, T.W., C. Garris, P. Jhingran, M. Bonafede, K.J. Tomaszewski, T. Bonus, R.M. Hahn and M. Schatz. 2010. Asthma That Is Not Well-Controlled Is Associated with Increased Healthcare Utilization and Decreased Quality of Life. J Asthma, in press.

Juniper, E.F., A. Chauhan, E. Neville, A. Chatterjee, K. Svensson, A.C. Mörk and E. Ståhl. 2004. Clinicians tend to overestimate improvements in asthma control: an unexpected observation. Prim Care Respir J 13: 181–4.

Juniper, E.F., J. Bousquet, L. Abetz and E.D. Bateman. 2006. Identifying 'well-controlled' and 'not well-controlled' asthma using the Asthma Control Questionnaire. Respir Med 100: 616–621.

Juniper, E.F., K. Svensson, A.C. Mörk and E. Ståhl. 2005. Measurement properties and interpretation of three shortened versions of the asthma control questionnaire. Respir Med 99: 553–8.

Katz, P.P., E.H. Yelin, M.D. Eisner and P.D. Blanc. 2002. Perceived control of asthma and quality of life among adults with asthma. Ann Allergy Asthma Immunol 89: 251–258.

Khalili, B., P.B. Boggs, R. Shi and S.L. Bahna. 2008. Discrepancy between clinical asthma control assessment tools and fractional exhaled nitricoxide. An Allergy Asthma Immunol 101: 124–129.

Lai, C.K., T.S. De Guia, Y.Y. Kim, S.H. Kuo, A. Mukhopadhyay, J.B. Soriano, P.L. Trung, N.S. Zhong, N. Zainudin and B.M. Zainudin; Asthma Insights and Reality in Asia-Pacific Steering Committee. 2003. Asthma control in the Asia-Pacific region: the Asthma Insights and Reality in Asia-Pacific Study. J Allergy Clin Immunol 111: 263–8.

Nathan, R.A., C.A. Sorkness, M. Kosinski, M. Schatz, J.T. Li, P. Marcus, J.J. Murray and T.B. Pendergraft. 2004. Development of the asthma control test: a survey for assessing asthma control. J Allergy ClinImmunol 113: 59–65.

O'Byrne, P.M., H.K. Reddel, G. Eriksson, O. Ostlund, S. Peterson, M.R. Sears, C. Jenkins, M. Humbert, R. Buhl, T.W. Harrison, S. Quirce and E.D. Bateman.

2010. Measuring asthma control: a comparison of three classification systems. Eur Respir J 36: 269–76.

O'Byrne, P.M., I.P. Naya, A. Kallen, D.S. Postma and P.J. Barnes. 2008. Increasing doses of inhaled corticosteroids compared to adding long-acting inhaled beta2-agonists in achieving asthma control. Chest 134: 1192–9.

Partridge, M.R., T. van der Molen, S.E. Myrseth and W.W. Busse. 2006. Attitudes and actions of asthma patients on regular maintenance therapy: the INSPIRE study. BMC Pulm Med 13; 6: 13.

Ponte, E.V., R. Franco, H.F. Nascimento, A. Souza-Machado, S. Cunha, M.L. Barreto, C. Naspitz and A.A. Cruz. 2008. Lack of control of severe asthma is associated with co-existence of moderate-to-severe rhinitis. Allergy 63: 564–9.

Rabe, K.F., M. Adachi, C.K. Lai, J.B. Soriano, P.A. Vermeire, K.B. Weiss and S.T. Weiss. 2004. Worldwide severity and control of asthma in children and adults: the global asthma insights and reality surveys. J Allergy Clin Immunol 114: 40–7.

Rabe, K.F., P.A. Vermeire, J.B. Soriano and W.C. Maier. 2000. Clinical management of asthma in 1999: the Asthma Insights and Reality in Europe (AIRE) study. Eur Respir J 16: 802–7.

Reddel, H.K., D.R. Taylor, E.D. Bateman, L.P. Boulet, H.A. Boushey, W.W. Busse, T.B. Casale, P. Chanez, P.L. Enright, P.G. Gibson, J.C. de Jongste, H.A. Kerstjens, S.C. Lazarus, M.L. Levy, P.M. O'Byrne, M.R. Partridge, I.D. Pavord, M.R. Sears, P.J. Sterk, S.W. Stoloff, S.D. Sullivan, S.J. Szefler, M.D. Thomas and S.E. Wenzel; American Thoracic Society/European Respiratory Society Task Force on Asthma Control and Exacerbations. 2009. An official American Thoracic Society/European Respiratory Society statement: asthma control and exacerbations: standardizing endpoints for clinical asthma trials and clinical practice. Am J Respir Crit Care Med 180: 59–99.

Skinner, E.A., G.B. Diette, P.J. Algatt-Bergstrom, T.T. Nguyen, R.D. Clark, L.E. Markson and A.W. Wu. 2004. The Asthma Therapy Assessment Questionnaire (ATAQ) for children and adolescents. Dis Manag 7: 305–13.

Thomas, M., S. Kay, J. Pike, A. Williams, J.R. Rosenzweig, E.V. Hillyer and D. Price. 2009. The Asthma Control Test (ACT) as a predictor of GINA guideline-defined asthma control: analysis of a multinational cross-sectional survey. Prim Care Respir J 18: 41–49.

4

Rhinosinusitis and Asthma

***Bachert Claus,*[1,a,*] *Gevaert Philippe,*[1,b]**
***Van Zele Thibaut*[1,c] and *Acke Frederic*[1,d]**

ABSTRACT

Chronic rhinosinusitis (CRS) is frequently associated with and has a clear impact on asthma. In this chapter, we describe the current knowledge in this common condition, including state of art diagnosis and treatment, results of recent research findings and insights into new perspectives.

CRS implies long-term inflammation of nasal and paranasal cavities. It is defined based on its symptoms, confirmed by endoscopy and/ or imaging. In fact, CRS is divided into two distinct entities due to its endoscopic appearance: with and without nasal polyps (CRSwNP and CRSsNP). While Th1 lymphocytes, IFN-γ, and TGF-β play an important role in CRSsNP, Th2 lymphocytes, IL-5 and a deficit in TGF-β are crucial in CRSwNP. Thus, remodeling and inflammatory patterns clearly differ in the disease entities. Moreover, *Staphylococcal* enterotoxins are able to function as superantigens and massively activate T-cells and B-cells in CRSwNP patients, resulting in polyclonal IgE formation. This activation of the immune system can give rise to other manifestations of inflammation, such as asthma. Indeed, CRSwNP and asthma frequently

[1]Department of Otorhinolaryngology (1P1), Ghent University Hospital, De Pintelaan 185, B-9000 Ghent, Belgium.
[a]Email: claus.bachert@ugent.be
[b]Email: philippe.gevaert@ugent.be
[c]Email: thibaut.vanzele@ugent.be
[d]Email: frederic.acke@ugent.be
*Corresponding author
List of abbreviations after the text.

co-occur and share many pathophysiological attributes, leading to the concept of united airways disease.

Treatment of CRS is based upon severity and includes symptomatic treatment, nasal corticosteroids, long-term antibiotics and/or a short course of oral corticosteroids. Surgery is an alternative when patients do not satisfactorily respond to medical treatment. Novel, promising treatment modalities are being developed thanks to the advances in CRS research.

INTRODUCTION

Chronic rhinosinusitis (CRS) is a common disease, affecting approximately 10% of the population, and a significant health problem. It does not only affect patient's health, directly by interfering symptoms and indirectly by influencing lower airway disease, it also has socioeconomic implications.

The impact on quality of life needs to be emphasized; questionnaires confirmed the effect of CRS on functioning in daily life, limitations due to symptoms, vitality, mental health. Patients indicate negative influences on work and sleep performance. Moreover, aggravation of comorbid conditions, such as asthma, does occur. Appropriate treatment of CRS leads to an improvement in quality of life (Fokkens et al. 2007).

Apart from this debilitating aspect, CRS also induces large economic costs for the society. Direct medical and surgical expenditures are high, and this condition also creates indirect costs by causing lost work and school days, as well as reduced productivity. Indeed, most CRS patients are of working age.

To correctly diagnose CRS and to set up a proper treatment is of utmost importance. This chapter aims to summarize the current state of knowledge about CRS regarding epidemiology, pathophysiology, diagnosis, association with asthma, and treatment.

DEFINITIONS OF CRS/NP

Sinusitis implies inflammation of the paranasal cavities or sinuses, but in practice almost always occurs with concomitant rhinitis.

Consequently, rhinosinusitis is a widely accepted term, used to describe an inflammatory process involving the mucosa of the nose and one or more sinuses. Nasal polyps are growths of this inflamed mucosal tissue, pedunculated with a slim or broad stalk or base.

Many tried to define these findings, resulting in differences in duration and severity, and variations due to technical possibilities. Because of the numerous guidelines, consensus documents and position papers, the European Academy of Allergology and Clinical Immunology (EAACI) composed an evidence based position paper about CRS and nasal polyposis (NP). The European Position Paper on Rhinosinusitis and Nasal Polyps (EP3OS) was published in 2005 and revised in 2007. A clinical definition of rhinosinusitis was proposed (Fokkens et al. 2007), which is internationally accepted nowadays. This definition states that the diagnosis of CRS/NP is based on symptoms and their duration, but also includes endoscopic and/or radiologic criteria (Table 1). Typical symptoms are nasal congestion and discharge, facial pressure and loss of smell; while symptoms must be present for at least 12 weeks to differentiate chronic from acute rhinosinusitis. In addition, endoscopic evidence of inflammation and/or computed tomography (CT) changes must be present.

CRS and NP are frequently seen as one entity, because it is hard to make a clear differentiation between them based on clinical criteria only. NP is often considered a subgroup of CRS, introducing the terms chronic rhinosinusitis with nasal polyps (CRSwNP) and chronic rhinosinusitis without nasal polyps (CRSsNP). In the case of CRSwNP, endoscopic evidence of polyp presence is required. However, new insights into the inflammatory pattern of CRSsNP and CRSwNP revealed differences in inflammatory and remodeling markers, suggesting two separate entities (Van Zele et al. 2006, Van Bruaene et al. 2009).

EPIDEMIOLOGY OF CRS/NP

The paucity of accurate epidemiologic data on CRS/NP makes an accurate estimate of the prevalence of CRSsNP and CRSwNP speculative. The prevalence of CRS/NP depends upon the selection of the study population, the diagnostic methods, and definitions used.

Table 1 EP3OS definition of rhinosinusitis, including nasal polyps.

Inflammation of the nose and the paranasal sinuses characterized by two or more symptoms (one of which should be either the first or the second)
- nasal blockage/obstruction/congestion
- nasal discharge (anterior/posterior nasal drip)
- facial pain/pressure
- reduction or loss of smell
and either
Endoscopic signs (one ore more of the following)
- polyps
- mucopurulent discharge primarily from middle meatus
- oedema/mucosal obstruction primarily in middle meatus
and/or
CT changes
- mucosal changes within the ostiomeatal complex and/or sinuses
Acute rhinosinusitis
< 12 weeks
complete resolution of symptoms
Chronic rhinosinusitis
> 12 weeks
without complete resolution of symptoms

The definition of rhinosinusitis including nasal polyps (Fokkens et al. 2007), states that the diagnosis is based on symptoms and their duration, but also includes endoscopic and/or radiologic criteria.
(with permission from Fokkens et al.)

In surveys the prevalence of CRS raises up to 16% (Fokkens et al. 2007). However, the prevalence of doctor-diagnosed CRS is much lower, a prevalence of 2% was found using ICD-9 codes and half of the diagnoses were confirmed by nasal endoscopy (Shashy et al. 2004). The prevalence rate of CRS is substantially higher in females and increases with age. Although there is an increased prevalence of allergic rhinitis in patients with CRS, the role of allergy in CRS remains unclear.

The prevalence of nasal polyps varies between 1 and 4% of the general population. Johansson et al. (2003) reported a prevalence of nasal polyps of 2.7% in the total Swedish population and a postal questionnaire in Finland revealed a prevalence of 4.3% (Hedman et al. 1999). However, cadaver dissections suggest a much higher prevalence of up to 42% (Johansson et al. 2003). Nasal polyps are clearly a disease of adulthood and are more frequent in men (2.2 to 1 ratio) (Hedman et al. 1999). The average onset is approximately at 42 yr (Fokkens et al. 2007) and the incidence increases with age. Under the age of 20 nasal polyps are uncommon and are mostly related to cystic fibrosis. The prevalence of allergy in patients with NP has been reported as varying from 10 to 54% and a clear relationship between allergy and NP has not been shown. On contrary asthma is clearly linked to NP with one-third of NP patients having a diagnosis of asthma. Especially late onset asthma is associated with the development of nasal polyps in 10–15% (Settipane and Chafee 1977). Samter's triad is a well-known clinical entity and is characterized by the triad of asthma, nasal polyps and intolerance to aspirin and aspirin-like medications. Of patients with aspirin sensitivity, 36–96% have nasal polyps and aspirin sensitivity is present in 5–10% of asthmatic patients (Samter and Beers 1968).

PATHOPHYSIOLOGY OF CRSsNP

Normal sinuses are free of bacterial growth. Cultures by sinus puncture from both adults and children with acute rhinosinusitis or exacerbations of CRS predominantly grow *Streptococcus pneumoniae*, *Haemophilus influenzae*, and *Moraxella catarrhalis*. In chronic maxillary sinusitis, anaerobic bacteria alone or mixed infections with facultative anaerobes and aerobes are predominant: *Staphylococcus aureus*, coagulase-negative *Staphylococcus*, *P. aeruginosa*, and anaerobes (Bachert and van Cauwenberge 2003). Studies show incredible variation, and it is unclear, whether bacterial infections truly contribute to CRS pathophysiology, or may just be secondary to the local milieu in an obstructed sinus encouraging bacterial growth. Recent discussions focus on the role of biofilms as nidus and defense mechanisms for germs; *Staphylococcus aureus* carrying biofilms have especially been found in recurrent CRS (Foreman and Wormald 2010).

In the sinus fluid of patients with CRS undergoing surgery, inflammatory cells are predominantly neutrophils, but a low percentage of eosinophils, mast cells, and basophils may also be found. In a study evaluating the percentage of tissue eosinophils, patients with untreated CRSsNP had less than 10% eosinophils (overall mean 2%), whereas in untreated NP specimen, samples showed more than 10% eosinophils (overall mean 50%) (Jankowski 1996). These observations suggest that tissue eosinophilia is not a hallmark of CRSsNP. The mucosal lining in CRS is characterized by basement membrane thickening, goblet cell hyperplasia, subepithelial edema, and mononuclear cell infiltration. An analysis of relevant cells and cytokines in CRS recently demonstrated activated T-cells and macrophages to be increased in number in CRSsNP compared to controls; in contrast to CRSwNP, no plasma cell accumulation could be found (Van Zele et al. 2006).

The characteristic cytokine pattern (Fig. 1) in CRSsNP consists of pro-inflammatory and neutrophil-associated cytokines (interleukin (IL)-1β, tumor necrosis factor (TNF)-α and IL-8) resulting in increased neutrophil presence and activation (myeloperoxydase levels); and of the T helper cell (Th)1-related interferon (IFN)-γ (Van Zele et al. 2006). Furthermore, CRSsNP is characterized by an elevated tumor growth factor (TGF)-β signaling, leading to an increased mucosal collagen deposition and fibrosis, a hallmark of chronic obstructive

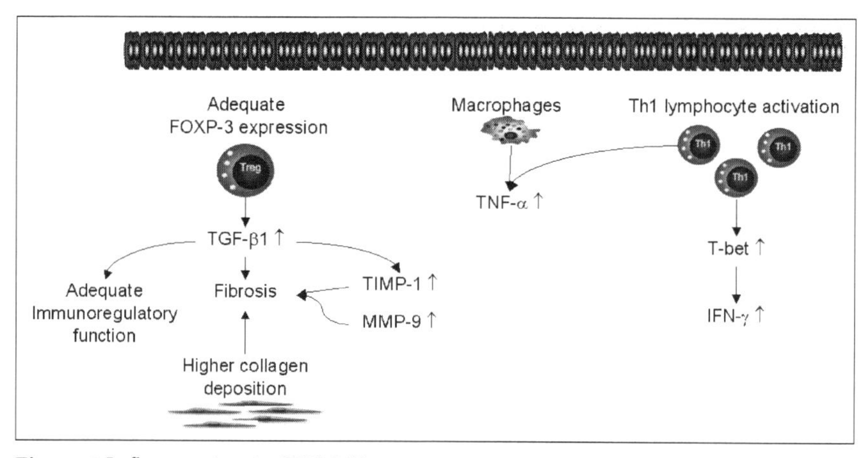

Figure 1 Inflammation in CRSsNP.
Different mediators play a role in the pathophysiology of CRSsNP, such as IFN-γ, TGF-β and TNF-α. (FOXP-3 = forkhead box P3 transcription factor, TIMP = tissue inhibitor metalloproteinase, MMP = matrix metalloproteinase, T-bet = T-box transcription factor Tbx21).
(unpublished figure of the authors)

sinus disease (Van Bruaene et al. 2009). Remodeling may actually precede inflammation, as recently shown in early CRS disease. The reason for the deregulation of TGF-β is unknown.

PATHOPHYSIOLOGY OF CRSwNP

In contrast to CRSsNP, CRSwNP presents with a deficit in TGF-β production with consecutive lack of collagen production and edema formation, and an imbalance of the metalloproteinases versus their tissue inhibitors with consecutive consumption of the extracellular matrix, resulting in the typical transparent berry-like protrusion of the mucosa from the middle meatus and the sinuses into the nasal cavity (Van Bruaene et al. 2009).

In CRSwNP, different types of inflammation may develop (Fig. 2) with a Th2 bias being frequent in Europe and the US, but a Th1/Th17 bias preferentially found in Asia (Zhang et al. 2008). IL-5 turned out to represent a key cytokine in the Th2 biased eosinophilic polyps; anti-IL-5 monoclonal antibodies, but not anti-IL-3 or anti-granulocyte macrophage colony-stimulating factor (GM-CSF), in vitro resulted in eosinophil apoptosis and decreased tissue eosinophilia (Simon et al. 1997). With IL-5, also IL-4 and IL-13 are over-expressed in CRSwNP, completing the Th2 biased environment by inducing immunoglobulin (Ig)E-synthesis and mucus hyperproduction. However, whereas in Europe 85% of CRSwNP patients do show a Th2 bias, 15% of those patients and the majority of patients in mainland China are IL-5 negative, and demonstrate a rather neutrophilic IFN-γ and IL-17 driven disease (Zhang et al. 2008). These differences may be decisive for future treatment approaches in CRSwNP disease using cytokine-specific humanized monoclonal antibodies.

Nasal polyps do harbor plenty of B-cells and plasma cells, which produce different types of immunoglobulins; a range of factors have recently been identified in polyp tissue which regulate immunoglobulin switch and synthesis. Specifically the production of IgE may be amplified by enterotoxins released from *S. aureus*, leading to a massive production of polyclonal IgE antibodies independent of the allergic status of the patient (Bachert et al. 2001). *Staphylococcal* enterotoxins function as superantigens, massively activating T-cells and B-cells in an antigen-independent way in mucosal

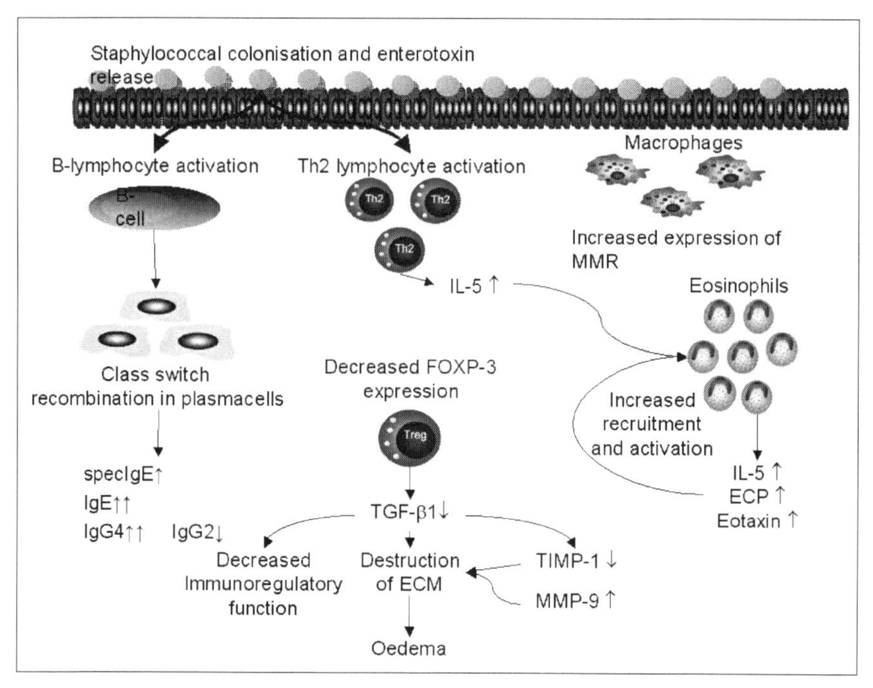

Figure 2 Inflammation in CRSwNP
Staphylococcal enterotoxins are involved in the pathophysiology of CRSwNP by functioning as superantigens and massively activating T-cells and B-cells. B-cells and plasma cells produce IgG, IgA and IgE antibodies, while Th2 lymphocytes predominate in European and US nasal polyps and produce IL-5 (FOXP-3 = forkhead box P3 transcription factor, TIMP = tissue inhibitor metalloproteinase, MMP = matrix metalloproteinase, ECM = extracellular matrix, MMR = macrophage mannose receptor).
(unpublished figure of the authors)

follicular structures (Gevaert et al. 2005). This massive activation of lymphocytes may result in the systemic spread of the inflammation, e.g., comorbid asthma. Also aspirin-exacerbated respiratory disease (AERD) can frequently be diagnosed in those patients, who typically show highly activated eosinophils, a prominent Th2 bias and polyclonal IgE formation, together with an up-regulation of cys-leucotrienes and a down-regulation of prostaglandin E4.

The rate of nasal colonization of *S. aureus* is significantly increased in polyp patients versus controls, and increases with the severity of airway disease, with a colonization rate of 88% in aspirin-sensitive asthmatic polyp patients (Van Zele et al. 2004); in up to 80% of those polyps, *staphylococcal* enterotoxin specific IgE antibodies

(SE-IgE) can be found locally. However, *S. aureus* was recently demonstrated to also reside intramucosally and intracellularly, and may form biofilms creating a nidus for recurrent infections and surviving mucosal defense mechanisms as well as antibiotic treatment (Foreman and Wormald 2010).

DIAGNOSTIC MANAGEMENT OF CRS/NP

Symptoms

CRSsNP and CRSwNP have similar signs and symptoms including post-nasal drip, rhinorrhea, nasal congestion and loss of smell. However the symptom pattern may vary. While CRSwNP is more often associated with loss of smell, in CRSsNP headache and facial pain are more prominent symptoms. However, facial pain alone does not constitute a suggestive history for rhinosinusitis in absence of another major symptom (Van Zele et al. 2006).

Nasal endoscopy

Anterior rhinoscopy may provide clues for the diagnosis of rhinosinusitis; however if performed without nasal endoscopy, it is insufficient to diagnose CRSsNP or CRSwNP. A nasal endoscopy should always be performed to examine the nasal cavity for edema, mucopurulent discharge, and nasal polyps especially at the middle meatus. In case of nasal polyps it has been shown that the patient's symptoms are not a good indicator for the size of polyps. Therefore the extent of the disease is preferably estimated by endoscopy using a staging system. Several staging systems for polyps have been proposed of which a scoring system with four steps proved to be the most reliable (Fokkens et al. 2007).

Imaging

The imaging modality of choice for evaluating the anatomical extent of nasal polyps is CT scanning. Of several CT staging systems the Lund Mackay has been validated in several studies; however a poor correlation between CT scores and symptoms scores has been described. Transillumination and plain sinus x-rays lack sensitivity and specificity and therefore are unreliable for the diagnosis of CRSsNP and CRSwNP. Magnetic resonance imaging (MRI) should be preferred if a neoplasm of the nasal cavity is suspected, brain structures are involved, or a fungal disease is expected (Fokkens et al. 2007).

Cytology

Cytology has not proved a useful tool in the diagnosis of chronic rhinosinusitis while a biopsy is indicated to exclude sinister and severe conditions such as neoplasm and vasculitis (Fokkens et al. 2007).

Blood testing

Eosinophilia in the peripheral blood is a hallmark of CRSwNP, particularly among Caucasians. Eosinophil numbers are significantly higher in nasal polyp tissue compared to CRSsNP and are further increased in patients with comorbid asthma and/or aspirin sensitivity, independent from atopy. In CRSwNP high total IgE levels in serum have been found especially in patients with comorbid asthma and aspirin intolerance.

Microbiology

The most common bacterial pathogens associated with CRS include coagulase-negative *staphylococci*, *S. aureus*, *viridians* group *streptococci*, gram-negative enteric rods (especially *Pseudomonas aeruginosa*), and anaerobes. Bacterial colonization and release of enterotoxins by *Staphylococcus aureus* may represent a crucial modifying factor in both upper and lower airway disease (Van Zele et al. 2004). *Pseudomonas aeruginosa* is commonly observed in immunocompromised patients and in cystic fibrosis. As *Pseudomonas* is mostly a nosocomial infection, it is frequently seen in recurrence after surgery and in more persistent disease.

CRS/NP AND ASTHMA (GENERAL CONSIDERATIONS)

The nose is essential in the protection and homeostasis of lower airways. Upper and lower airways are not only allied anatomically, they are both lined with a pseudo-stratified respiratory epithelium and equipped with an arsenal of innate and acquired immune defense mechanisms (Fokkens et al. 2007). Epidemiologically, up to 60% of CRS patients have lower airway involvement, mostly hidden small airway disease but also 24% asthma (Ragab et al. 2004). Asthma and chronic obstructive pulmonary disease (COPD) are frequently found in CRSsNP patients. However, CRSwNP is accounting for the largest part of this high prevalence; approximately one-third of nasal polyp patients have a diagnosis of asthma, while 7% of asthmatics have nasal polyps (Bachert et al. 2006). Comorbid asthma in nasal polyp patients frequently can be classified as severe and treatment resistant; AERD is specifically frequent in these patients. In patients with concomitant asthma, more severe and persistent forms of sinus disease seem to occur. Moreover, medical treatment and surgery of the upper airways in asthmatics positively influence the course of asthma. Several studies showed that, when optimal treatment is given to patients suffering from CRSsNP or CRSwNP, bronchial symptoms decrease and medication use for asthma is reduced (Fokkens et al. 2007).

This relationship gave rise to the concept of 'united airways disease'. Proposed mechanisms of interplay between the upper and lower airway include pharyngobronchial reflexes, aspiration of inflammatory mediators from the nose into the bronchial tree, and systemic inflammatory response secondary to either rhinosinusitis or asthma (Bachert et al. 2006). When studying the latter, cytokine patterns in sinus tissue of CRS highly resemble those of bronchial tissue in asthma, explaining the presence of eosinophils in both conditions. The strong relationship between CRSwNP and asthma could partly be explained by the association with aspirin-exacerbated respiratory disease. The role of *S. aureus* and the presence of SE-IgE in the pathogenesis of asthma merits further investigation, as this could account for the link with CRSwNP (Bachert et al. 2010, Kowalski et al. 2011).

The concept of united airways will be of clinical relevance, because new treatment modalities affecting both upper and lower airway inflammation are being developed.

SEVERE COMMON AIRWAY DISEASE

A cluster analysis of factors present in nasal polyps revealed that SE-IgE, total IgE and eosinophilic cationic protein, a marker of eosinophil activation, are all associated with comorbid asthma (Bachert et al. 2010). As described, *staphylococcal* superantigens may induce polyclonal IgE formation and eosinophil activation, suggesting that superantigens may orchestrate systemically disseminating severe airway disease; SE-IgE can be frequently found also in the sera of these patients. Furthermore, factors such as IFN-γ could also be identified which have a protective effect.

SE-IgE serves as a marker of superantigen impact, and indicates a polyclonal B-cell activation; high total IgE versus low specific IgE concentrations are a typical finding then, with specific IgE forming a small fraction of the total IgE (less than 2%). Studies in human polyp tissue clearly demonstrated the functionality of specific IgE antibodies in the absence of a positive skin test or specific IgE in the serum of the patient (Zhang et al. 2011). The presence of thousands of specific IgE antibodies directed to inhalant and mucosal allergens, including enterotoxins, in the tissue could give rise to a continuous stimulation of immune cells and thus contribute to the severity and persistence of the inflammation. Consequently, anti-IgE treatment studies have been set up as proof-of-concept.

Similar mechanisms have recently been shown to be relevant in severe lower airway disease independent of an upper airway equivalent. Typical risk factors for severe refractory asthma are female gender, history of wheezing in childhood, presence of hypersensitivity to aspirin and BMI; in addition, the mean concentrations of SE-IgE were three-fold higher in patients with severe refractory asthma as compared to patients with non severe asthma or controls (Kowalski et al. 2011). The presence of SE-IgE in about 80% of severe asthmatics carried a significant risk to have serum total IgE levels above 100 kU/l. Furthermore, increased concentrations of SE-IgE antibodies were significantly associated with impaired respiratory function parameters and increased airway reversibility. About one-third of the severe asthma patients suffered from AERD, and about 50% of them had a history of CRSwNP. These findings argue for a role of *staphylococcal* enterotoxins in the pathogenesis of severe upper and lower airway disease, including asthma.

CURRENT AND FUTURE CONCEPTS OF TREATMENT

Corticosteroids

Corticosteroids are able to improve upper (CRSsNP and CRSwNP) and lower (asthma) airway inflammatory diseases by potent anti-inflammatory activity. Long-term treatment with nasal spray reduces inflammation and polyp size, and improves nasal symptoms. Consequently, intranasal corticosteroids are the keystone of CRS treatment. Oral corticosteroids, restricted to short courses because of numerous side effects, are also able to improve the clinical course of NP; however their effects, in particular symptom improvement, are short-lasting (Van Zele et al. 2010; Fig. 3). Repeated courses of oral corticosteroids may lead to systemic side effects.

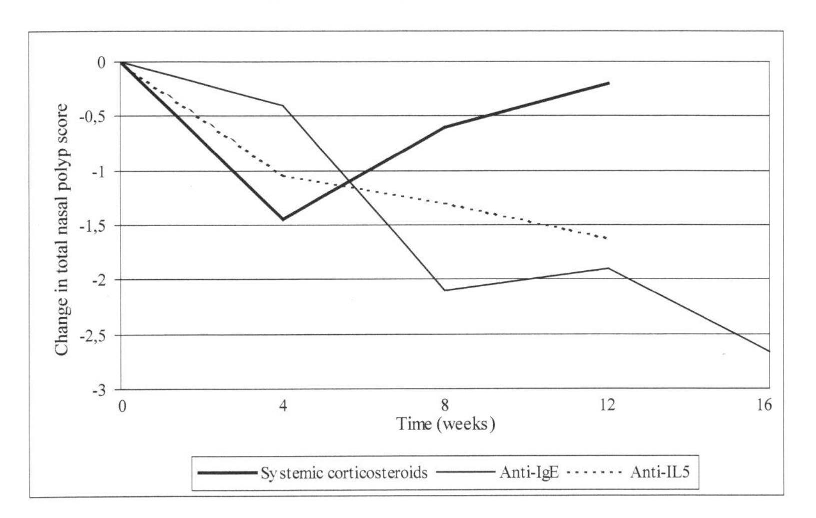

Figure 3 Effect of anti-IgE and anti-IL-5 treatment on nasal polyp size compared to systemic corticosteroids
The change in nasal polyp size (sum of 4-point scale results of both nostrils) is shown for systemic corticosteroids, anti-IgE and anti-IL-5 treatment. The group treated with systemic corticosteroids (methylprednisolone 32mg/d on days 1–5, 16mg/d on days 6–10, and 8mg/d on days 11–20) experienced a large, but short-lasting effect on nasal polyp size (Van Zele et al. 2010). The anti-IgE (4–8 subcutaneous injections of omalizumab over 16 weeks, maximum dose of 375mg) and anti-IL-5 (2 intravenous injections of 750mg mepolizumab over 4 weeks) results are the findings of two randomized controlled trials in patients with severe nasal polyposis (unpublished data of the authors). These groups experienced a large effect on nasal polyp size as well, and moreover it was maintained over several months.
(unpublished figure of the authors).

Antibiotics

The effect of short-term antibiotic treatment in CRS and exacerbations of CRS is unclear and needs further investigation (Fokkens et al. 2007). Increasing attention is paid to long-term antibiotic treatment of CRS. The effect of low-dose antibiotics, such as macrolides and tetracyclines is not only attributable to their antibacterial activity, but also to their anti-inflammatory effect. Particularly patients with disease exacerbated by the *Staphylococcus aureus* enterotoxin could benefit from this treatment (Van Zele et al. 2010).

Other Medical Treatment Modalities

Standard conservative treatment for CRS is based upon topical steroids and long-term antibiotics (Fig. 4 and Fig. 5). Many other types of preparations have been investigated, but substantial evidence for

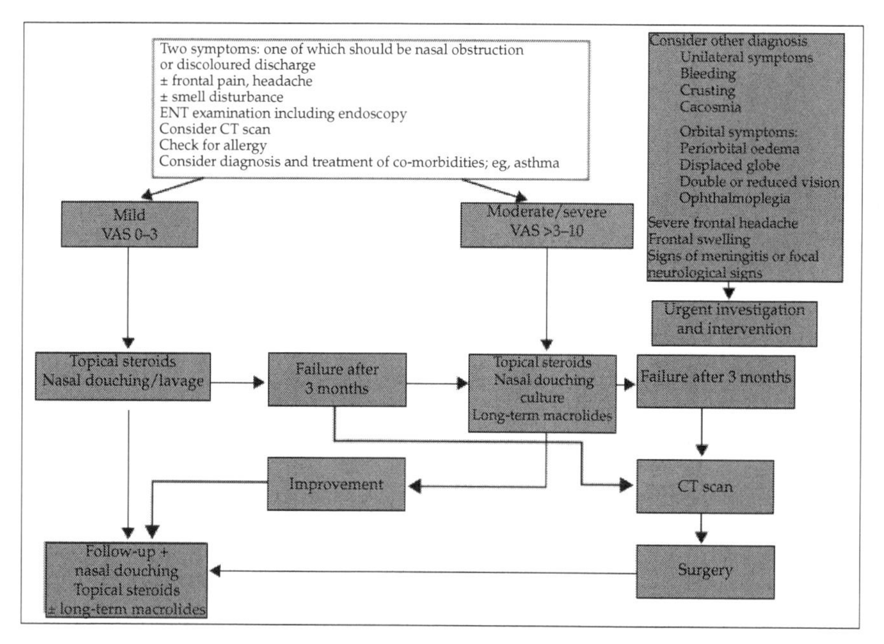

Figure 4 Treatment scheme for adults with CRSsNP.
This treatment scheme, developed for otorhinolaryngologists (Fokkens et al. 2007), depicts the treatment steps in CRSsNP. Treatment depends upon severity (VAS = visual analogue scale of patient-assessed symptom severity).
(with permission from Fokkens et al.)

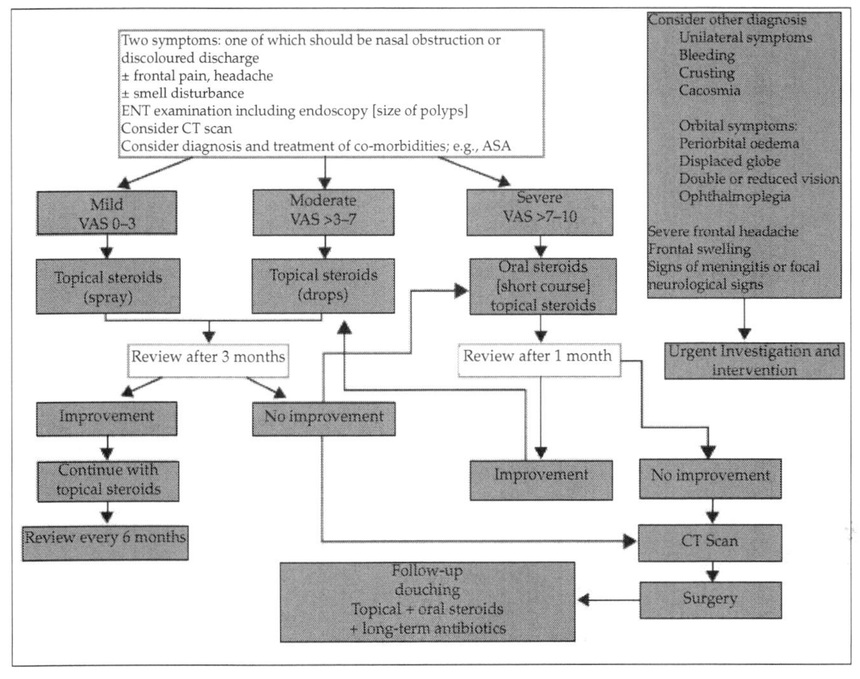

Figure 5 Treatment scheme for adults with CRSwNP.
This treatment scheme, developed for otorhinolaryngologists (Fokkens et al. 2007), depicts the treatment steps in CRSwNP. Treatment depends upon severity (VAS = visual analogue scale of patient-assessed symptom severity).
(with permission from Fokkens et al.)

their benefit is poor (Fokkens et al. 2007). Symptomatic medications include nasal douching/lavage, mucolytics, antihistamines and decongestants, which could achieve limited relieve of symptoms. The use of antimycotics in CRS is controversial (Ebbens et al. 2006). Antileukotrienes are a potentially effective treatment of CRSwNP, but need further investigation. Aspirin desensitization may be implicated in protection against CRS with nasal polyp recurrence (Fokkens et al. 2007).

Surgery

A large review of literature with highly consistent results suggests that patients with CRSsNP and CRSwNP benefit from sinus surgery (Fokkens et al. 2007). Functional endoscopic sinus surgery (FESS) techniques are standard today; nasal polyps need to be removed completely. In the majority of CRS patients, appropriate medical treatment is as effective as surgical treatment short-term. Sinus surgery should thus be reserved for patients who do not satisfactorily respond to medical treatment. Major complications occur in less than 1%, and revision surgery is performed in approximately 10% within 3 yr.

Future Concepts of Treatment

As the pathophysiology of CRSsNP and CRSwNP receives more attention, mediator-based treatments are being developed. Inhibiting IL-5 by intravenous administration of monoclonal antibodies is proven to be safe and effective in CRSwNP patients (Gevaert et al. 2006). Moreover, anti-IgE has proven beneficial effects on CRSwNP in patients treated for refractory severe asthma (Vennera et al. 2010). Studies to evaluate the effect of anti-IL-5 and anti-IgE in CRSwNP patients in a long-term perspective are promising (Fig. 3).

DIFFERENTIAL DIAGNOSIS

Wegener

Wegener's granulomatosis is a relatively common systemic disease that affects the upper airways in 90%. Patients with Wegener's granulomatosis normally present with symptoms of a prolonged upper airway tract infection, unresponsive to antibiotics. There is a typical pain over the dorsum with sero-sanguineous to purulent nasal discharge, friable mucosa with crusts, and nasal ulcers. Patients have a typical bony and cartilage destruction resulting in saddle nose deformities. Orbital masses and subglottic disease may occur. Careful biopsy of affected septal mucosa and turbinates, combined with positive c-anti-neutrophil cytoplasm autoantibodies (c-ANCA) findings are essential in the diagnostic work-up (Erickson and Hwang 2007).

Sarcoidosis

Sarcoidosis is a multisystem granulomatous disorder of unknown etiology that typically affects young adults and mostly manifests in the lungs. It is characterized pathologically by the presence of non-caseating granulomas. Upper respiratory tract sarcoidosis may involve the larynx, pharynx, nares and sinuses (Reed et al. 2010). Involvement of the paranasal sinuses is common among patients with intranasal sarcoidosis with symptoms of nasal obstruction, nasal crusting and anosmia. In a subgroup epistaxis and nasal polyposis can occur. Non-caseating granulomas are seen on biopsy of the nasal mucosa.

Churg Strauss Syndrome

The Churg Strauss syndrome is an eosinophil rich granulomatous vasculitis and has been associated with nasal polyp formation, allergic rhinitis, asthma and peripheral blood and tissue eosinophilia. The exact etiology is unknown; however there is a heightened Th1 and Th2 function with increased eosinophil recruitment and decreased eosinophil apoptosis (Hellmich et al. 2003). n-ANCA are detected in about 40 to 60% of patients. The mucosal destruction is less prominent as compared to Wegener's granulomatosis. Often nasal polyps may occur in the prodromal phase of the disease many years before the development of chronic eosinophilic gastroenteritis and pneumonia.

Cystic Fibrosis (CF)

The majority of cystic fibrosis patients develop sinus disease. Ten tot 32% have nasal polyps on endoscopic examination; however only 10% of CF patients actively report sinonasal symptoms (Robertson et al. 2008). Although polyps are clinically identical with their adult counterparts, CF polyps are generally more neutrophilic and not eosinophilic in nature (Van Zele et al. 2006). On CT scan, sinonasal bulging or displacement of the lateral nasal wall and demineralisation of the uncinate process are typical signs of CF patients. Hypoplasia or aplasia of the paranasal sinuses has been uniformly reported in patients with CF.

Primary Ciliary Dyskinesia

Patients with Kartagener's syndrome and primary ciliary dyskinesia have a long history of respiratory infection and rhinosinusitis is a common problem. Most patients develop chronic sinusitis, involving the maxillary, ethmoidal, and frontal sinuses, although the frontal sinuses often fail to develop (Leigh et al. 2009).

Practice and Procedures

Scheme for adults with CRSsNP or CRSwNP for ENT-specialists (Fokkens et al. 2007):

Diagnosis

- Symptoms (present longer than 12 weeks): two or more symptoms, one of which should be eiter nasal blockage/ obstruction/congestion or nasal discharge (anterior/posterior nasal drip), ± facial pain/pressure, ± reduction or loss of smell.
- Signs: ENT examination, endoscopy; review primary care physician's diagnosis and treatment; questionnaire for allergy and if positive, allergy testing if not already been done.

Treatment: Decide on severity of symptomatology using VAS (Fig. 4 for CRSsNP and Fig. 5 for CRSwNP).

KEY FACTS OF THE CONCEPT OF UNITED AIRWAYS DISEASE

- United airways disease is a concept, suggesting upper and lower airways diseases are associated, e.g., allergic rhinitis and asthma, nasal polyposis and asthma.
- The nose and lungs are linked anatomically, have the same covering tissue and use similar immune defense mechanisms.

- Asthma and CRS, especially CRSwNP, co-occur more than their distinct prevalences would suggest.
- Many theories have been proposed to explain this link, of which the systemic inflammatory response secondary to either rhinosinusitis or asthma, is most supported.
- Recently, attention is paid to the role of *staphylococcal* enterotoxins in the pathogenesis of severe upper and lower airway disease.
- Treating rhinosinusitis in asthmatics positively influence the course of asthma.
- New treatment modalities affecting both upper and lower airway inflammation are being developed.

SUMMARY POINTS

- Chronic rhinosinusitis is a frequent condition that can be subdivided into CRS with or without nasal polyps.
- Different inflammatory mediators predominate in CRSwNP compared to CRSsNP.
- The diagnosis of CRS is made by symptom report and nasal endoscopy and/or CT scan imaging.
- There is a link between CRSwNP and asthma, suggesting the concept of 'united airways disease', with common pathophysiological mechanisms.
- Severe persistent CRSwNP disease with comorbid asthma is associated with elevated total IgE and specific IgE antibodies to *staphylococcal* enterotoxins.
- Treatment of CRS includes long-term nasal corticosteroids and eventually symptomatic treatment. The administration of long-term antibiotics or a short course of oral corticosteroids can be considered.
- Surgery, particularly FESS, is an alternative when patients do not satisfactorily respond to medical treatment.

DEFINITIONS AND EXPLANATION OF WORDS AND TERMS

Asthma: a disease characterized by recurrent attacks of breathlessness and wheezing due to inflammation of the air passages in the lungs.

Chronic rhinosinusitis: a condition with inflammation of nose and sinuses, in which symptoms persist for more than 12 weeks. It can be subdivided into chronic rhinosinusitis with or without nasal polyps.

Functional endoscopic sinus surgery: surgery performed through the nose, using a rigid endoscope. It is the most common type of surgery for chronic rhinosinusitis with and without nasal polyps, and aims to restore function and to relieve symptoms.

Nasal endoscopy: the use of a thin endoscope put into the nose to provide a clear view of the nasal structures and nasopharynx.

Nasal polyposis: a mucosal inflammatory condition in which grape-like edematous swellings (nasal polyps) grow in the nasal and paranasal cavities.

Samter's triad: the combination of asthma, nasal polyps and intolerance to aspirin and aspirin-like medications.

Staphylococcal enterotoxin specific IgE antibodies: proteins produced by the immune system of the body in a reaction induced by toxins or superantigens released by *Staphylococcus aureus.*

Staphylococcus aureus: a gram-positive bacterium frequently living on the skin or nose. It can be present without giving symptoms, but it is a frequent cause of infection too.

LIST OF ABBREVIATIONS

AERD : aspirin-exacerbated respiratory disease
ANCA : anti-neutrophil cytoplasm autoantibodies
CF : cystic fibrosis
CRS : chronic rhinosinusitis
CRSsNP : chronic rhinosinusitis without nasal polyps

CRSwNP	:	chronic rhinosinusitis with nasal polyps
CT	:	computed tomography
IFN	:	interferon
Ig	:	immunoglobulin
IL	:	interleukin
NP	:	nasal polyposis
SE-IgE	:	*staphylococcal* enterotoxin specific immunoglobulin E antibodies
TGF	:	tumor growth factor
Th	:	T helper cell
TNF	:	tumor necrosis factor

REFERENCES

Bachert, C., P. Gevaert, G. Holtappels, S.G. Johansson and P. van Cauwenberge. 2001. Total and specific IgE in nasal polyps is related to local eosinophilic inflammation. J Allergy Clin Immunol 107: 607–614.

Bachert, C. and P. van Cauwenberge. Nasal polyps and sinusitis. pp. 1421–1436. *In:* N.F. Adkinson, J.W. Yunginger, W.W. Busse, B.S. Bochner, S.T. Holgate and F.E.R. Simons. [eds.] 2003. Middleton's Allergy: principles and practice. Mosby, St. Louis, USA.

Bachert, C., J. Patou and P. Van Cauwenberge. The role of sinus disease in asthma. 2006. Curr Opin Allergy Clin Immunol 6: 29–36.

Bachert, C., N. Zhang, G. Holtappels, L. De Lobel, P. van Cauwenberge, S. Liu, P. Lin, J. Bousquet and K. Van Steen. 2010. Presence of IL-5 protein and IgE antibodies to staphylococcal enterotoxins in nasal polyps is associated with comorbid asthma. J Allergy Clin Immunol 126: 962–968.

Ebbens, F.A., G.K. Scadding, L. Badia, P.W. Hellings, M. Jorissen, J. Mullol, A. Cardesin, C. Bachert, T.P. van Zele, M.G. Dijkgraaf, V. Lund and W.J. Fokkens. 2006. Amphotericin B nasal lavages: not a solution for patients with chronic rhinosinusitis. J Allergy Clin Immunol 118: 1149–1156.

Erickson, V.R. and P.H. Hwang. 2007. Wegener's granulomatosis: current trends in diagnosis and management. Curr Opin Otolaryngol Head Neck Surg 15: 170–176.

Fokkens, W., V. Lund and J. Mullol. 2007. European position paper on rhinosinusitis and nasal polyps 2007. Rhinol Suppl 20: 1–136.

Foreman, A. and P.J. Wormald. 2010. Different biofilms, different disease? A clinical outcomes study. Laryngoscope 120: 1701–1706.

Gevaert, P., G. Holtappels, S.G. Johansson, C. Cuvelier, P. van Cauwenberge and C. Bachert. 2005. Organisation of secondary lymphoid tissue and local IgE formation to Staphylococcus aureus enterotoxins in nasal polyp tissue. Allergy 60: 71–79.

Gevaert, P., D. Lang-Loidolt, A. Lackner, H. Stammberger, H. Staudinger, T. Van Zele, G. Holtappels, J. Tavernier, P. van Cauwenberge and C. Bachert. 2006.

Nasal IL-5 levels determine the response to anti-IL-5 treatment in patients with nasal polyps. J Allergy Clin Immunol 118: 1133–1141.
Hedman, J., J. Kaprio, T. Poussa and M.M. Nieminen. 1999. Prevalence of asthma, aspirin intolerance, nasal polyposis and chronic obstructive pulmonary disease in a population-based study. Int J Epidemiol 28: 717–722.
Hellmich, B., S. Ehlers, E. Csernok and W.L. Gross. 2003. Update on the pathogenesis of Churg-Strauss syndrome. Clin Exp Rheumatol 21: S69–S77.
Jankowski, R. 1996. Eosinophils in the pathophysiology of nasal polyposis. Acta Otolaryngol 116: 160–163.
Johansson, L., A. Akerlund, K. Holmberg, I. Melén and M. Bende. 2003. Prevalence of nasal polyps in adults: the Skövde population-based study. Ann Otol Rhinol Laryngol 112: 625–629.
Kowalski, M.L., M. Cieślak, C.A. Pérez-Novo, J.S. Makowska and C. Bachert. 2011. Clinical and immunological determinants of severe/refractory asthma (SRA): association with Staphylococcal superantigen-specific IgE antibodies. Allergy 66: 32–38.
Leigh, M.W., J.E. Pittman, J.L. Carson, T.W. Ferkol, S.D. Dell, S.D. Davis, M.R. Knowles and M.A. Zariwala. 2009. Clinical and genetic aspects of primary ciliary dyskinesia/Kartagener syndrome. Genet Med 11: 473–487.
Ragab, A., P. Clement and W. Vincken. 2004. Objective assessment of lower airway involvement in chronic rhinosinusitis. Am J Rhinol 18: 15–21.
Reed, J., R.D. deShazo, T.T. Houle, S. Stringer, L. Wright and J.S. Moak 3rd. 2010. Clinical features of sarcoid rhinosinusitis. Am J Med 123: 856–862.
Robertson, J.M., E.M. Friedman and B.K. Rubin. 2008. Nasal and sinus disease in cystic fibrosis. Paediatr Respir Rev 9: 213–219.
Samter, M. and R.F. Beers Jr. 1968. Intolerance to aspirin. Clinical studies and consideration of its pathogenesis. Ann Intern Med 68: 975–983.
Settipane, G.A. and F.H. Chafee. 1977. Nasal polyps in asthma and rhinitis. A review of 6,037 patients. J Allergy Clin Immunol 59: 17–21.
Shashy, R.G., E.J. Moore and A. Weaver. 2004. Prevalence of the chronic sinusitis diagnosis in Olmsted County, Minnesota. Arch Otolaryngol Head Neck Surg 130: 320–323.
Simon, H.U., S. Yousefi, C. Schranz, A. Schapowal, C. Bachert and K. Blaser. 1997. Direct demonstration of delayed eosinophil apoptosis as a mechanism causing tissue eosinophilia J Immunol 158: 3902–3908.
Van Bruaene, N., L. Derycke, C.A. Perez-Novo, P. Gevaert, G. Holtappels, N. De Ruyck, C. Cuvelier, P. Van Cauwenberge and C. Bachert. 2009. TGF-beta signaling and collagen deposition in chronic rhinosinusitis. J Allergy Clin Immunol 124: 253–259.
Van Zele, T., P. Gevaert, J.B. Watelet, G. Claeys, G. Holtappels, C. Claeys, P. van Cauwenberge and C. Bachert. 2004. Staphylococcus aureus colonization and IgE antibody formation to enterotoxins is increased in nasal polyposis. J Allergy Clin Immunol 114: 981–983.
Van Zele, T., S. Claeys, P. Gevaert, G. Van Maele, G. Holtappels, P. Van Cauwenberge and C. Bachert. 2006. Differentiation of chronic sinus diseases by measurement of inflammatory mediators. Allergy 61: 1280–1289.
Van Zele, T., P. Gevaert, G. Holtappels, A. Beule, P.J. Wormald, S. Mayr, G. Hens, P. Hellings, F.A. Ebbens, W. Fokkens, P. Van Cauwenberge and C. Bachert. 2010.

Oral steroids and doxycycline: two different approaches to treat nasal polyps. J Allergy Clin Immunol 125: 1069–1076.
Vennera, M.D., C. Picado, J. Mullol, I. Alobid and M. Bernal-Sprekelsen. 2010. Efficacy of omalizumab in the treatment of nasal polyps. Thorax (in press).
Zhang, N., T. Van Zele, C. Perez-Novo, N. Van Bruaene, G. Holtappels, N. DeRuyck, P. Van Cauwenberge and C. Bachert. 2008. Different types of T-effector cells orchestrate mucosal inflammation in chronic sinus disease. J Allergy Clin Immunol 122: 961–968.
Zhang, N., G. Holtappels, P. Gevaert, J. Patou, B. Dhaliwal, H. Gould and C. Bachert. 2011. Mucosal tissue polyclonal IgE is functional in response to allergen and SEB. Allergy 66: 141–148.

5

Serious Psychological Distress and Asthma

Michael E. King

ABSTRACT

Effective management of asthma requires a regimen of care that may be compromised by psychological distress. Clinical studies have linked anxiety and depression to functional impairment, reduced health-related quality of life, increased risk of hospitalization, and a decreased ability to manage symptoms among persons with asthma. This chapter reviews recent research on serious psychological distress (SPD) among persons with asthma and discusses implications for asthma care and public health practice. Over the last 10 years, nine population-based studies have reported estimates of SPD in the general population and among persons with asthma. Surveys conducted in Australia, Canada, and the United States confirmed that SPD is common among persons with asthma and has a negative impact on functional ability, quality of life, and health. In primary care settings, the use of brief measures, like the Kessler non-specific psychological distress scale, may raise awareness of mental health (MH) needs and identify potential barriers to asthma management. Likewise, increased involvement of specialists and education about psychological distress can help reduce negative stigma, facilitate the early identification and treatment of mental illness, and promote MH. From a public health perspective, integrating MH and

Air Pollution and Respiratory Health Branch, Division of Environmental Hazards and Health Effects, National Center for Environmental Health, Centers for Disease Control and Prevention, 4770 Buford Highway, MS F-58, Chamblee, GA 30341; Email: nzk7@cdc.gov

List of abbreviations after the text.

asthma control activities may reduce healthcare utilization, morbidity, and mortality in populations with co-occurring asthma and SPD.

INTRODUCTION

According to the World Health Organization (WHO), more than 450 million people around the world have a mental disorder (WHO 2010a). People diagnosed with chronic mental disorders frequently report multiple other chronic physical conditions and the link between respiratory and mental health (MH) has been well-documented (Muehrer 2002). Numerous clinical and community-based studies have reported rates of psychological distress among asthma patients that are twice those of the general population (Goodwin 2003). This relationship is important because comorbid mental disorders can increase the severity of asthma symptoms, is associated with more risk behaviors, and present a serious barrier to asthma care (Strine et al. 2008). The purpose of this chapter is to: (a) identify population-based studies of asthma and serious psychological distress (SPD), (b) describe the prevalence of psychological distress among those with asthma, and (c) discuss potential implications for health and public health practice. Although evidence presented in this chapter is cross-sectional, limiting determination of causality or the directionality of associations, this summary of patterns of prevalence and factors associated with co-occurring SPD and asthma may serve as a useful resource for future research and applied public health practice.

MAIN TEXT

Asthma is among the most common chronic respiratory diseases, affecting over 300 million people worldwide (WHO 2010b). Globally, the prevalence of asthma has increased for over 30 years, resulting in a total cost associated with medical care and productivity loss of $300 to $1,300 per patient and an estimated 250,000 deaths annually (GINA 2010). These figures highlight the substantial health and economic burden of asthma, much of which has been attributed to inadequate symptom control, and underscore the need to identify populations at increased risk for negative asthma-related health outcomes.

Although there is no cure for asthma, the availability of effective management strategies suggests that asthma control is a realistic treatment goal for most people. Symptoms can be prevented entirely or their severity reduced with routine access to medication, medical care, appropriate self-management, and environmental precautions to reduce exposure to things that make asthma worse (GINA 2010). However, this regimen relies on strategies that may be compromised by psychological distress, making effective asthma control a challenge for persons diagnosed with a mental disorder.

What is Psychological Distress?

Studies of asthma are characterized by a diversity of psychological outcomes, ranging from negative emotions, like sadness or fear, to anxiety and depression (Goodwin 2003, Zielinski and Brown 2003). Terminology aside, any experience of psychological distress may be acute and transient or more severe and chronic. To be diagnosed as a psychological disorder, established Diagnostic and Statistical Manual of Mental Disorders-IV-Revised (DSM-IV-TR) criteria must be met and symptoms must have persisted for a specified duration, limiting major life activities (American Psychiatric Association 2000). Signs and symptoms of select disorders common among patients with asthma are provided in Table 1 (Goodwin 2003, Zielinski and Brown 2003).

Previous reviews have summarized research on co-occurring asthma, anxiety, and depression, but it is important to note that much of the evidence was derived from clinic-based (e.g., treatment-seeking) or convenience samples, which limits the generalizability of findings to other populations (Goodwin 2003, Zielinski and Brown 2003). Epidemiologic research is necessary to measure the prevalence and determinants of asthma, psychological distress, and answer questions about the impact of distress on health outcomes. Measuring psychological outcomes in general population surveys is challenging, however, because diagnostic interviews are lengthy, difficult to administer, and it is often impractical to ask respondents directly about mental illness because many do not have access to psychiatric care or are unaware that they have mental illness (Sekula et al. 2003). To facilitate epidemiologic measurement, brief screening tools based on DSM-IV-TR criteria have been developed. For this chapter, serious psychological distress (SPD) was defined as a non-specific category of distress characterized by symptoms of anxiety or depression, as measured by the Kessler scale (Kessler et al. 2002).

Table 1 Diagnostic and Statistical Manual of Mental Disorders-IV Signs and Symptoms for Generalized Anxiety Disorder, Panic Disorder, and Major Depressive Disorder.

Generalized Anxiety Disorder	**Major Depressive Disorder**
Anxiety and worry associated with at least three of the following six symptoms. Symptoms must: • be present for more days than not over a period of 6 months	Five or more of the nine symptoms listed. At least one of those must be either depressed mood or loss of interest. Symptoms must: • be present most of the time nearly every day for a period of 2 or more weeks • cause clinically significant distress or impairment in social, occupational, or other important areas of functioning • not be related to bereavement • not meet criteria for a mixed episode, defined as the presence of both depressed and manic symptoms
Signs and Symptoms	
• Restlessness or feeling keyed up or on edge	
• Being easily fatigued	
• Difficulty concentrating or mind going blank	
• Irritability	**Signs and Symptoms**
• Muscle tension	• Depressed or sad mood
• Sleep disturbance	• Markedly diminished interest or pleasure in all or almost all activities, especially those the individual normally enjoys
Panic Disorder	
Recurrent, unexpected, panic attacks with at least one attack followed by 1 month or more of the following symptoms. Symptoms must: • not be due to physiologic effects of a substance (drug-use or prescribed medication) or medical condition • not be better accounted for by another mental disorder	• Sleep disturbance • Feeling of worthlessness or excessive/inappropriate guilt • Fatigue or loss of energy • Indecisiveness or diminished ability to think or concentrate • Psychomotor agitation nearly every day
Signs and Symptoms	• A change in appetite, or a significant weight change when not intentionally trying to lose or gain weight
• Persistent concern about having additional panic attacks	

Table 1 contd....

Table 1 contd....

Generalized Anxiety Disorder	Major Depressive Disorder
• Worry about the implications of the attack or its consequences • A significant change in behavior related to the attacks	• Recurrent thoughts of death (not just fear of dying), suicidal ideation without a specific plan, a specific plan for committing suicide, or a suicide attempt
Adapted from the American Psychiatric Association. 2000. Diagnostic and Statistical Manual of Mental Disorders, 4th ed, text rev. Washington, DC, American Psychiatric Association.	

The clinical diagnosis of psychological disorders is based on the presence and duration of signs and symptoms listed in the Diagnostic and Statistical Manual of Mental Disorders-IV-Revised. Examples of disorders that have been identified among patients with asthma include major depression, generalized anxiety, and panic disorder. (Original material of author; adapted from the American Psychiatric Association, 2000).

The Kessler Scale

The Kessler non-specific psychological distress scale is a brief, dimensional measure designed to identify persons with a high likelihood of having a diagnosable serious mental illness in general population surveys. Both the Kessler 10 (K 10), and the shorter Kessler 6 (K 6), ask about the frequency of symptoms of psychological distress during the past 30 days (Table 2); details of scale development, psychometric properties, and scoring have been described elsewhere (Kessler et al. 2002). The Kessler scale is currently included in national surveys in Australia, Canada, the U.S., in the WHO-administered World Mental Health Surveys, and has been translated into 11 languages other than English (National Comorbidity Survey 2011).

Literature Search

The Ovid portal to Medline, PsychInfo, and EMBASE was searched using the key words: psychological-distress and asthma. Studies were selected for inclusion if they: (a) indicated asthma was an outcome,

Table 2 Kessler non-specific psychological distress scales.

<table>
<tr><td colspan="2">The Kessler scales ask about feelings of non-specific psychological distress during the past 30 days. The K 6 is a truncated version of the Kessler 10 and responses to both scales are rated on a 5-point Likert-type scale, ranging from "None of the time" to "All of the time." Scoring of individual items is typically based on a scale of between 0 and 4 points, according to increased frequency of the problem. Points are totaled across all items; assuming an identical scoring approach for either scale, this would result in a total K 6 score ranging from 0 to 24 and a K 10 score ranging from 0 to 40.[1]</td></tr>
<tr><td colspan="2">Example response scale and cut-offs [2]</td></tr>
<tr><td>Response = Points
All of the time (1) = 4 points
Most of the time (2) = 3 points
Some of the time (3) = 2 points
A little of the time (4) = 1 point
None of the time (5) = 0 points</td><td>Cutoffs applied to the general population
K-10 [3] | | K-6
0–5 | No Distress | 0–7
6–19 | Mild Distress | 8–12
20–40 | Serious Distress | 13+</td></tr>
<tr><td>K 6 items[4]</td><td>K 10 additional items [4]</td></tr>
<tr><td>During the past 30 days, about how often did you feel:
....nervous?
...hopeless?
...restless or fidgety?
...so depressed that nothing could cheer you up?
...that everything was an effort?
...worthless?</td><td>*all K 6 items*
+
...tired out for no good reason?
...so nervous that nothing could calm you down?
...so restless you could not sit still?
...so sad that nothing could cheer you up?</td></tr>
<tr><td colspan="2">Source: National Comorbidity Survey 2011
1: Alternative scoring methods exist for the K 10
2: Different cut-off values, specific to certain populations/settings have been published for the K 10 and K 6
3: See Schmitz et al. 2009
4: Item ordering not specific to the K6 or K10</td></tr>
</table>

The Kessler 10 (K 10) and Kessler 6 (K 6) scales are brief, validated measures designed to screen general populations for serious mental illness using the fewest items possible. Both scales measure symptoms of anxiety and depression experienced in the past 4 weeks. High scores on the K 10/K 6 are indicative of serious psychological distress and a possible need for mental health services. (Original material of author; adapted from the National Comorbidity Survey 2011).

(b) estimated SPD, using the K 10 or K 6, and (c) reported an analysis of a large, population-based sample. Results were limited to English language articles of high key-word relevance (based on search-engine options) published between 2000 and 2011. The search, completed on March 1, 2011, produced 54 unique citations; of these, a total of nine reports provided estimates of the prevalence of SPD and asthma. These studies are summarized in Table 3 and their implications for health and public health practice are discussed below.

Serious Psychological Distress and Asthma

Prevalence

From 2000–2011, nine published reports described prevalence estimates for asthma, SPD, and the co-occurrence of both conditions (Table 3). The prevalence of asthma ranged from 6.6% to 11.2% for current asthma and 10.9% to 13.2% for lifetime asthma. The prevalence of SPD ranged from 2.0% to 10.8% in the general population. Prevalence estimates of the co-occurrence of SPD and asthma ranged from 3.6% to 24.4% (Table 3). Socio-demographic factors that characterize populations with asthma and SPD included: being female, of working age (between 25–64 years), of Hispanic ethnicity, being unemployed or unable to work, of low socioeconomic status, being a current smoker, and being uninsured (McVeigh et al. 2006, Pratt et al. 2007, Oraka et al. 2010).

Health outcomes

Disability. Effective asthma management requires the ability manage symptoms, adhere to treatment plans, and implement strategies to avoid things in the environment that make asthma worse. Therefore, it is not surprising that level of disability was a primary outcome associated with SPD and asthma (Table 3). The Australian Bureau of Statistics (2006) reported that persons with asthma experienced more psychological distress, rated their health lower and had more days away from work or school, compared to those without asthma. Although Pratt et al. (2007) did not examine disability status among those with asthma specifically, the presence of SPD was associated

Table 3 Population-based studies of asthma and serious psychological distress (SPD), 2000–2011.

Study	Setting	Survey (Year)	Sample	Prevalence			Key Finding(s)
				Asthma Only	SPD[1] Only	SPD and Asthma	
Adams et al. 2004	Australia	Health and Wellbeing Survey (2000)	7,443 adults aged 18 years or older	11.2%[2]	2.7%[3]	3.6% [4]	Psychological distress and decreased feelings of control were common & associated with worse perceived physical health status among people with asthma.
Ampon et al. 2005	Australia	National Health Survey (2001)	14,641 adults aged 18-64 years	11.2%[2]	12.3%[3]	20.8%	The presence of asthma contributed to worse quality of life and accounted for more disability than diabetes, but less than arthritis.
Australian Bureau of Statistics 2006	Australia	National Health Survey (2004–2005)	Not reported	10.0%[2]	2.0%[3]	5.0% [5]	The proportion of people with SPD and asthma was more than double that of SPD and no asthma.
McVeigh et al. 2006	United States (New York City)	Community Health Survey (2002 & 2003)	9,342 & 9,599 adults aged 18 or older	Not reported	4.8%[6]	10.9% [7]	The age-adjusted prevalence of psychological distress among people with asthma & diabetes was double that of those without either condition.
Pratt et al. 2007	United States	National Health Interview Survey (2001–2004)	123,610 adults aged 18 years or older	Not reported	10.8%[6]	24.4% [8]	Adults with SPD were more likely to be obese, current smokers, require help with daily activities, use more medical services.
Schmitz et al. 2009	Canada	Canadian Community and Health Survey (2005)	62,274 household residents aged 12 years or older	6.6%[2]	2.1%[3]	5.6% [9]	The prevalence of functional disability, activity limitation, and poor health was higher among those with SPD and asthma.

McGuire et al. 2009	United States	Behavioral Risk Factor Surveillance System (2007)	35,845 adults aged 65 or older	10.9%[10]	2.7%[6]	19.9%	Older adults had a lower prevalence of SPD compared to younger age groups. Those with SPD had higher prevalence of lifetime asthma, compared to no SPD, and more comorbid chronic conditions.
McKnight-Eily et al. 2009	United States	Behavioral Risk Factor Surveillance System (2007)	220,448 adults aged 18 years or older	13.2%[10]	4.0%[6]	7.0%	Adults with SPD and asthma reported more days with activity limitation than those without SPD.
Oraka et al. 2010	United States	National Health Interview Survey (2001–2007)	186,738 adults aged 18 or older	7.0%[2]	3.0%[6]	7.5%	A negative association between SPD and quality of life was found. Adults with asthma and SPD had lower socioeconomic status, history of smoking or alcohol use, and more chronic comorbid illnesses.

1. *SPD = Serious Psychological Distress.*
2. *Current doctor diagnosed asthma.*
3. *Measured using the Kessler 10 (K10).*
4. *"High Risk" SPD among those with asthma based on a K10 score of 30–50.*
5. *"Very High" distress among adults aged 18 or older.*
6. *Measured using the Kessler 6 (K6).*
7. *Age-adjusted prevalence of SPD among those reporting "asthma and diabetes".*
8. *SPD among those reporting lifetime diagnosis of chronic lung disease, including asthma or chronic bronchitis.*
9. *"High distress" among those with asthma based on a K10 score of 19–40.*
10. *Doctor diagnosed asthma during lifetime.*

From 2000–2011, nine published studies from Australia, Canada, and the United States reported estimates of asthma and psychological distress from large-scale, population-based surveys that confirm serious psychological distress (SPD) is common among populations with asthma.(Original material of author).

with a history of asthma or chronic bronchitis and several disabilities, including needing assistance with activities of daily living, having vision and hearing impairment, difficulty walking, and difficulty getting out to go shopping or to social events. Similarly, McKnight-Eily et al. (2009) found that a significantly higher proportion of adults with SPD and asthma reported activity limitation in the past month, compared to those with asthma but no SPD (adjusted odds ratio 4.78, 95% confidence interval 3.70–6.18). More recently, a study based on the 2005 Canadian Community and Health Survey (CCHS) found that the prevalence of disability and activity limitation was significantly higher in respondents who reported asthma and co-occurring SPD, compared to those with asthma or SPD alone, and that the prevalence of disability-days increased with the severity of SPD (Schmitz et al. 2009).

Health-related quality of life. Three studies measured the association between comorbid asthma and SPD on health-related quality of life (HRQOL) using validated or multi-item measures. Adams et al. (2004) found adults with asthma and SPD had significantly lower scores on the physical component of the SF-12 quality of life scale compared to those with either asthma or SPD alone. In contrast, among those with SPD, the mental health summary scores of adults with asthma did not differ from those without asthma. Other factors that had a significant association with quality of life included (a) feeling a lack of control over health, (b) age, (c) financial status, (d) education, and (e) number of days with work limitations. Ampon et al. (2005) assessed HRQOL using a combination of four measures (e.g., life satisfaction, health status, SPD, reduced activity days) and reported that the presence of asthma accounted for 3–8% of people reporting adverse quality of life outcomes; this proportion was higher than that attributed to diabetes and lower than that attributed to arthritis. The most recent study by Oraka et al. (2010) measured HRQOL using a composite measure of perceived health and activity limitations called the Health and Activities Limitations Index (HALex). A negative association between SPD and HRQOL was found for all respondents, regardless of asthma status, with quality decreasing as severity of distress increased. Notably, overall HRQOL was worse for those with asthma, compared to those without asthma, while SPD was a stronger predictor of worse HRQOL, compared to asthma alone.

Chronic conditions and risk behaviors. Most of the studies in this review reported that populations with asthma and SPD had more other chronic health conditions and health risk behaviors. For example, Oraka et al. (2010) found that adults with asthma who were not working, current smokers, and had three or more comorbid conditions had significantly higher odds of SPD, while McKnight-Eily et al. (2009) noted that as the number of additional chronic conditions increased, so did the degree of perceived activity limitation. McGuire et al. (2009) found that adults aged 65 years and older with SPD had a higher prevalence of lifetime asthma ($p < 0.01$) and were more likely to report at least two chronic health conditions ($p < 0.01$), fair to poor health, to have at least 14 physically or mentally unhealthy days in the past month, and be smokers. Two other studies reported a significant association between SPD and potentially-modifiable risk factors such as smoking and a sedentary lifestyle or obesity (McVeigh et al. 2006, Pratt et al. 2007).

Discussion: SPD and Asthma

Evidence from recent population-based studies confirms that SPD is more prevalent among persons with asthma, compared to the general population. Specifically, surveys conducted in Australia, Canada, and the U.S. over the last 10 years found that the prevalence of SPD among those with asthma was more than twice that among those without asthma (Fig. 1). These estimates were similar to those reported for more specific mental disorders among adults with asthma, where prevalence has ranged from 3.8% for anxiety disorder to 22.2% for panic disorder and 30.6% for lifetime diagnosis of major depression (Goodwin 2003, Strine et al. 2008). Conversely, population-based estimates of SPD appear to be lower than those of from clinic-based samples, where structured diagnostic interviews have found rates of 13% for social phobia, 16% for panic disorder, and 25% for major depression among asthma patients (Brown et al. 2000). Higher prevalence rates in clinical samples may be due to the possibility that patient populations have a greater level of illness overall, or it may be that the Kessler scale only captures the most severe cases of distress and that mild cases of distress remain undetected. Regardless, SPD is clearly prevalent among populations with asthma and, as a valid marker for more specific mental disorders, is likely to

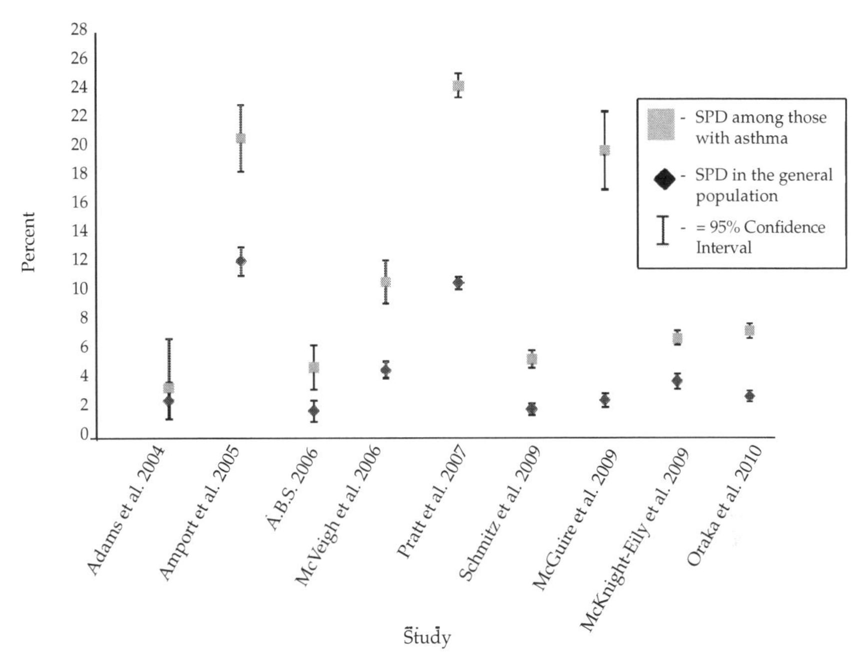

Figure 1 Prevalence of serious psychological distress (SPD) among persons with asthma and the general population by study.
From 2000–2011, nine reports from population-based surveys conducted in Australia, Canada, and the United States used the Kessler non-specific psychological distress scale to estimate serious psychological distress (SPD). Persons with asthma consistently report higher levels of SPD compared to persons in the general population. *Squares* indicate SPD among those with asthma, *diamonds* indicate SPD in the general population, and *vertical bars* indicate 95% confidence intervals. (Original material of author).

be associated with adverse asthma outcomes correlated with anxiety or depression, including increased severity of airway obstruction, functional impairment, risk of hospitalization, and a decreased ability to manage symptoms.

In addition to documenting prevalence, epidemiologic studies support previous clinic and community-based evidence that describes multiple ways that SPD might make asthma worse and complicate treatment. A theoretical model illustrating select cognitive and behavioral factors is presented in Fig. 2. Co-occurring asthma and SPD is associated with high levels of physical, psychological, and social disability at the population level; this is consistent with previous studies indicating patients with anxiety/depression have

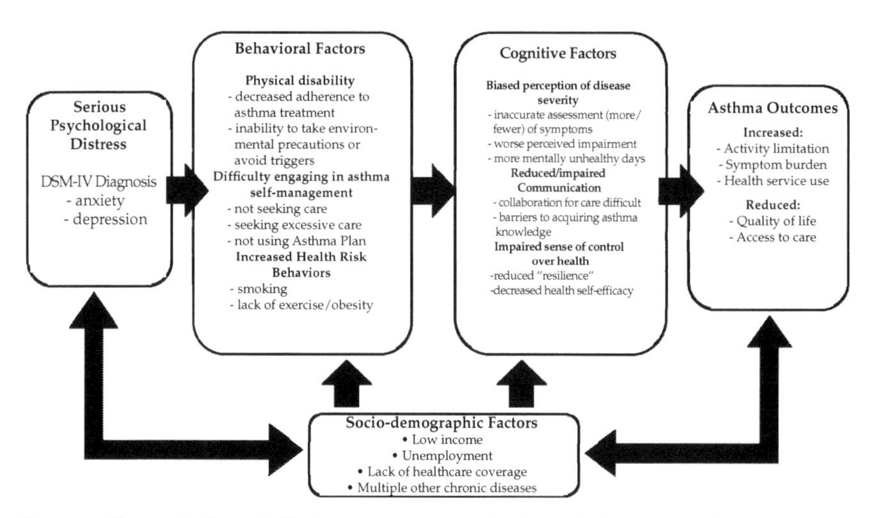

Figure 2 Theoretical model linking serious psychological distress to adverse asthma outcomes.
Epidemiologic studies of asthma, together with evidence from clinic and community-based studies, suggest that the presence of psychological distress can adversely impact asthma health outcomes via cognitive and behavioral factors. (Original material of author; adapted from Katon et al. 2004).

difficulty adhering to treatment plans and medication regimens (Lavoie et al. 2006). Perceptions of poor health and lack of control over physical and mental health issues reported by populations with SPD reflect findings that patients with mental disorders have trouble assessing asthma symptoms objectively and managing exacerbations, delay seeking care or seek care unnecessarily, and have a reduced sense of asthma self-efficacy (ten Brinke et al. 2001, Katon et al. 2004). Across studies, a similar pattern of socio-demographic factors (i.e., smoking, poverty, lack of insurance) has been found to predict both SPD and comorbid SPD and asthma. This pattern is consistent with longitudinal evidence suggesting the association between asthma and mental disorders may not be causal, but rather is a reflection of common genetic, behavioral, or social factors that increase individual susceptibility to respiratory and mental disorders alike (Goodwin et al. 2004). Although a complete review of potential mechanisms is beyond the scope of this chapter, consideration of factors presented in this model may facilitate the identification of research gaps.

Overall, populations with both asthma and SPD report higher levels of disability and worse HRQOL compared to persons with either condition alone and those with no chronic health conditions.

Persons with asthma and SPD also report more other chronic health conditions and risk behaviors compared to those without SPD. Considering the level of disability associated with SPD and asthma, it is surprising that no published studies reported healthcare utilization as an outcome. Hospitalizations and emergency department (ED) visits are indicative of serious episodes of asthma that could be avoided with proper asthma management. While Adams et al. (2004) noted that healthcare utilization was an outcome of the Health and Well-being Survey in Australia, findings specific to asthma were not included in the published report. Likewise, although Pratt et al. (2007) reported that persons with SPD used more medical services, including doctor visits and mental health professional visits, compared to persons without SPD, this finding was not stratified by asthma status. In contrast, a recent U.S. study associated depression and anxiety with high levels of disability, activity limitation, and decreased asthma control, including more days with symptoms and visits to the doctor and ED, among adults with asthma (Strine et al. 2008). Although the directionality of cause and effect between SPD and adverse health outcomes remains unclear (e.g., severe asthma symptoms might cause SPD or vice versa), taken together, these findings indicate that psychological distress may present a serious barrier to asthma care and be an important predictor of adverse asthma outcomes.

PRINCIPLES AND PRACTICES

Implications for Asthma Care

Most cases of asthma are diagnosed and managed in primary care settings, so it is particularly important for clinicians to be aware that SPD occurs more often than would be expected by chance in populations with asthma and this can adversely affect treatment outcomes (GINA 2010). Best-practice guidelines recommend the use of validated instruments for the screening and assessment for psychological factors, including depression, anxiety, and HRQOL, particularly when determining a patient's ability to manage their disease (GINA 2010). The use of brief measures like the Kessler scale in primary care settings can raise awareness of the need for MH care among patients and providers and facilitate the early identification and treatment of mental disorders.

Populations with co-occurring asthma and SPD report high levels of disability and activity limitation, indicating that psychological distress may adversely impact a patient's ability to manage symptoms, interfere with routine access to care, and result in increased healthcare utilization. Therefore, it is possible that increased involvement of specialists in primary asthma care settings might help improve access to and coordination of psychiatric care. Patients with low levels of psychological distress could benefit from the addition of stress-reduction and coping strategies to existing asthma self-management instruction, while clinicians should consider referring patients with SPD for more comprehensive diagnostic testing. In outpatient settings, the introduction of specialist asthma nurses has been found to improve care and reduce costs (Kamps et al. 2004), while referral to certified asthma educators and additional education about psychological issues may improve application of evidence-based guidelines and reduce negative stigma associated with mental disorders.

Despite growing awareness of the relevance of psychological distress, the needs of persons with asthma and SPD remain complex. Persons with SPD tend to report more other chronic diseases, in addition to risk behaviors like smoking, alcohol use, and physical inactivity, while the use of certain asthma medications, including beta-agonists, can cause symptoms similar to panic disorders. Furthermore, many people with SPD never seek care and an increasing number report unmet MH care needs related to cost or lack of insurance coverage, suggesting access to care may not be keeping pace with increasing awareness (The Commonwealth Fund 2011).

Implications for Public Health Practice

The aim of a public health approach to asthma is to decrease disease burden and improve health outcomes for large groups of people through effective, affordable interventions. A draft logic model for how MH might be integrated with a public health approach to asthma is presented in Fig. 3. A first step, at the population level, might be to use national and local surveillance to increase awareness of the burden of SPD and asthma among the public, healthcare providers, and policy makers (WHO 2010b). Another step would be to strengthen asthma healthcare through partnerships that support treatment

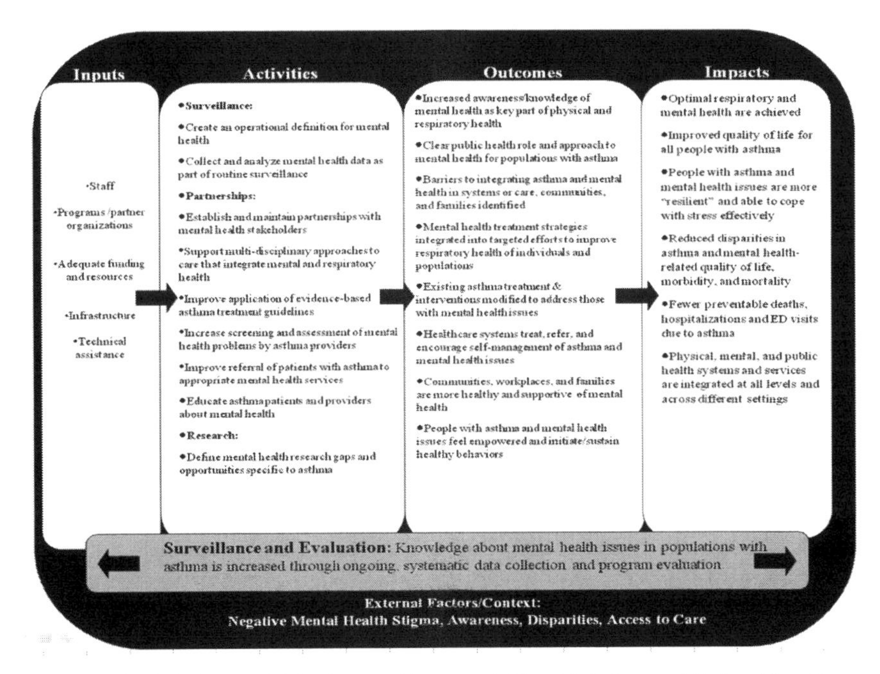

Figure 3 Draft logic model for integrating mental health into a public health asthma control program.
Integrating mental health into public health approaches to asthma control—including public health surveillance, partnerships, and research—presents an opportunity to leverage limited resources, target groups at increased risk, and achieve optimal respiratory and mental health at the population level. (Original material of author; adapted from Lando et al. 2006).

found to be effective for those with mental disorders (WHO 2010). Studies in this chapter indicate that SPD is an important contributor to the burden of asthma and health-related disability, quality of life, and risk behaviors. This suggests that the needs of patients with asthma and SPD may best be addressed through multidisciplinary, multi-factorial approaches that leverage limited resources to address different diseases and health behaviors concurrently. For example, chronic disease self-management education programs that help teach people to manage symptoms of chronic illness, adhere to treatment and medication recommendations, and improve functional ability have been found effective for participants with depression (Jerant et al. 2008) and asthma (Tousman et al. 2007). Likewise, promoting the application of evidence-based guidelines for asthma care and supporting public policies that facilitate access to affordable healthcare may help improve respiratory health, reduce costs, and possibly prevent unnecessary deaths due to asthma (Goodwin 2003).

Limitations

The limitations of evidence in this chapter should be considered. First, the studies included in Table 3 were cross-sectional in design, which precludes drawing any conclusions about causality or the directionality of the association between SPD and asthma. Second, study outcomes were assessed by self-report, potentially resulting in inaccurate, or biased, estimates of the disease burden. Third, the Kessler scale is more specific than sensitive (e.g., those identified are likely to meet diagnostic criteria for serious mental illness, but many could have been missed) and may have underestimated SPD prevalence. Fourth, the characteristics of the populations from which the survey samples were drawn are unique to each country; because SPD and asthma are influenced by socio-demographic factors, the validity of conclusions drawn from these findings are limited to the country in which each study was conducted. Finally, this review focused on published studies of SPD and asthma, omitting other potentially relevant reports that did not appear in electronic databases.

CONCLUSION

While many questions remain unanswered about the relationship between psychological distress and asthma, it is clear that our understanding of the magnitude and determinants of SPD has improved over the last 10 years. Because SPD is common among persons with asthma, it may be useful to include validated screening instruments when assessing patients with asthma. Studies confirm that SPD has a substantial and negative impact on asthma-related health outcomes, and is associated with more chronic health conditions and risk behaviors, underscoring the importance of approaches to healthcare that promote mental health as a key component of respiratory health.

Disclaimer: The findings and conclusions in this report are those of the author and do not necessarily represent the official position of the Centers for Disease Control and Prevention.

KEY FACTS

- The Kessler scale measures non-specific symptoms of psychological distress in the past 30 days.
- Two versions of the scale exist: the Kessler 10 (K10) and a briefer version, the Kessler 6 (K6).
- The Kessler scale was developed in 1992 to measure "Serious Mental Illness" (defined as: the presence of a 12 month DSM-IV disorder plus functional impairment) but was found to capture less severe cases in 2004 and has since been used to measure "Serious Psychological Distress."
- Kessler scale scores strongly discriminate between cases and non-cases of anxiety and depression in the community, with higher scores reflecting more serious distress and a high likelihood of having a diagnosable DSM-IV disorder.
- The Kessler scale has been used in Australia, Canada, the U.S., and in the World Health Organization's World Mental Health Initiative.

SUMMARY POINTS

- Asthma self-management strategies may be adversely impacted by the presence of psychological distress.
- Anxiety and depression are common among asthma patients and have been associated with worse asthma outcomes.
- Brief scales—like the Kessler scale—have been developed to screen for mental disorders when time is limited or personnel trained to administer diagnostic interviews are absent.
- Evidence from nine population-based surveys in Australia, Canada, and the United States confirm that SPD is twice as prevalent among people with asthma, compared to the general population.
- Across studies, more severe psychological distress was associated with worse HRQOL, regardless of asthma.

- Primary care settings present an opportunity to screen for SPD, educate patients and providers, and reduce negative stigma associated with mental disorders.
- Integrating mental health with population-based approaches to respiratory health is an important public health priority.
- Healthcare approaches that address the needs of persons with multiple chronic conditions, including mental disorders, may help reduce costs and improve outcomes.

DEFINITIONS AND EXPLANATION OF WORDS AND TERMS

Serious psychological distress (SPD): Serious psychological distress is a non-specific indicator of symptoms of anxiety or depression that would be likely to meet diagnostic criteria for a serious mental illness in the last 30 days, as measured by the Kessler non-specific psychological distress scale.

Asthma: Asthma is a prevalent chronic inflammatory disorder of the airways characterized by recurring episodes of wheezing, breathlessness, chest tightness, or coughing and associated with reversible airflow obstruction. Notably, in survey research, a distinction is often made between persons reporting ever receiving a diagnosis of asthma ("Lifetime Asthma") and those reporting both a diagnosis and current symptoms of asthma ("Current Asthma").

Surveillance: The systematic collection, analysis, and dissemination of health-related survey data from large populations.

Primary care: Basic health care provided as an initial approach to treatment.

Disability: A physical or mental limitation in activity resulting from a health condition.

Population-based: In the context of this chapter, "population-based" refers to surveys or information drawn from a large, defined group of people with distinct socio-demographic characteristics.

Health-related quality of life: A conceptual measure of physical, mental, emotional, and/or social functional ability related specifically to health status.

Health risk behaviors: Behaviors that may negatively impact health, such as smoking, physical inactivity, or alcohol consumption.

Asthma self-management/control: The ability to respond to or prevent asthma symptoms via appropriate use of medication, access to medical care, and environmental precautions that reduce exposure to things that make asthma worse.

Prevalence: The total number of persons with a particular disease in a population at a given time.

LIST OF ABBREVIATIONS

SPD	:	Serious Psychological Distress
MH	:	Mental Health
K 6	:	Kessler 6 Non-Specific Psychological Distress Scale
K 10	:	Kessler 10 Non-Specific Psychological Distress Scale
WHO	:	World Health Organization
DSM-IV-TR	:	Diagnostic and Statistical Manual-IV-Text Revision
HRQOL	:	Health-Related Quality of Life
ED	:	Emergency Department
SF-12	:	Short Form Health Survey (12 item version)
HALex	:	The Health and Activities Limitations Index

REFERENCES

Adams, R.J., D.H. Wilson, A.W. Taylor, A. Daly, E. Tursan d-Espaignet, E. Dal Grande and R.E. Ruffin. 2004. Psychological factors and asthma quality of life: a population based study. Thorax 59: 930–935.

American Psychiatric Association. 2000. Diagnostic and Statistical Manual of Mental Disorders (Revised 4th ed.). Washington, DC: American Psychiatric Association.

Ampon, R.D., M. Williamson, P.K. Correll and G.B. Marks. 2005. Impact of asthma on self-reported health status and quality of life: a population based study of Australians aged 18–64. Thorax 60: 735–739.

Australian Bureau of Statistics. 2006. Asthma in Australia: a snapshot, 2004–05. Retrieved February 2, 2011 from: http: //www.abs.gov.au/ausstats/abs@.nsf/mf/4819.0.55.001/.

Brown, E.S., D.A. Khan and S. Mahadi. 2000. Psychiatric diagnoses in inner-city outpatients with moderate to severe asthma. International Journal of Psychiatry in Medicine 30: 319–327.

Global Initiative for Asthma (GINA). 2010. GINA Report, Global Strategy for Asthma Management and Prevention. Retrieved March 1, 2011 from: http: // www.ginasthma.com.

Goodwin, R.D. Asthma and anxiety disorders. pp. 51–71. *In:* E. S. Brown [ed.] 2003. Asthma: Social and Psychological Factors and Psychosomatic Syndromes. Advances in Psychosomatic Medicine. Volume 24. Basel, Karger.

Goodwin, R.D., D.M. Fergusson and L.J. Horwood. 2004. Asthma and depressive and anxiety disorders among young persons in the community. Psychological Medicine 34: 1465–1474.

Jerant, A., R. Kravit, M. Moore-Hill and P. Franks. 2008. Depressive symptoms moderated the effect of chronic illness self-management training on self-efficacy. Medical Care 46: 523–531.

Kamps, A.W.A., R.J. Roorda, J.L.L. Kimpen, A.W. Overgoor-van de Groes, L.C.J.A.M. van Helsdingen-Peek and P.L.P Brand. 2004. Impact of nurse-led outpatient management of children with asthma on healthcare resource utilization and costs. European Respiratory Journal 23: 304–309.

Katon, W.J., L. Richardson, P. Lozano and E. McCauley. 2004. The relationship of asthma and anxiety disorders. Psychosomatic Medicine 66: 349–355.

Kessler, R.C., G. Andrews, L.J. Colpe, E. Hiripi, D.K. Mroczek, S.L.T. Normand, E.E. Walters and A.M. Zaslavsky. 2002. Short screening scales to monitor population prevalences and trends in non-specific psychological distress. Psychological Medicine 32: 959–976.

Lando, J., S.M. Williams, B. Williams and S. Sturgis. 2006. A logic model for the integration of mental health into chronic disease prevention and health promotion. Preventing Chronic Disease: Public Health Research, Practice, and Policy 3: 1–4.

Lavoie, K.L., S.L. Bacon, S. Barone, A. Cartier, B. Ditto and M. Labrecque. 2006. What is worse for asthma control and quality of life: depressive disorders, anxiety disorders or both? Chest 130: 1039–1047.

McGuire, L.C., T.W. Strine, S. Vachirasudlekha, L.A. Anderson, J.T. Berry and A.H. Mokdad. 2009. Modifiable characteristics of a healthy lifestyle and chronic health conditions in older adults with or without serious psychological distress, 2007 Behavioral Risk Factor Surveillance System. International Journal of Public Health 54: S84–S93.

McKnight-Eily, L.R., L.D. Elam-Evans, T.W. Strine, M.M. Zack, G.S. Perry, L. Presley-Cantrell, V.J. Edwards, and J.B. Croft. 2009. Activity limitation, chronic disease, and comorbid serious psychological distress in US adults—BRFSS. 2007. International Journal of Public Health 54: S111–S119.

McVeigh, K.H., S. Galea, L.E. Thorpe, C. Maulsby, K. Henning and L.I. Sederer. 2006. The epidemiology of nonspecific psychological distress in New York City, 2002 and 2003. Journal of Urban Health: Bulletin of the New York Academy of Medicine 83: 394–405.

Muehrer, P. 2002. Research on co-morbidity, contextual barriers, and stigma: an introduction to the special issue. Journal of Psychosomatic Research 53: 843–845.

National Comorbidity Survey. K 10 and K 6 Scales. 2011. Retrieved February 12, 2011 from: http: //www.hcp.med.harvard.edu/ncs/k6_scales.php .

Oraka, E., M.E. King and D.B. Callahan. 2010. Asthma and serious psychological distress: prevalence and risk factors among US adults, 2001–2007. Chest 137: 609–616.

Pratt, L.A., A.N. Dey and A.J. Cohen. 2007. Characteristics of adults with serious psychological distress as measured by the K6 scale: United States, 2001–04. Advance data from vital and health statistics; no 382. Hyattsville, MD: National Center for Health Statistics.

Schmitz, N., J. Wang, A. Malla and A. Lesage. 2009. The impact of psychological distress on functional disability in asthma: results from the Canadian Community Health Survey. Psychosomatics 50: 42–49.

Sekula, L.K., J. DeSantis and V. Gianetti. 2003. Considerations in the management of the patient with comorbid depression and anxiety. Journal of the American Academy of Nurse Practitioners 15: 23–33.

Strine, T.W., A.H. Mokdad, L.S. Balluz, J.T. Berry and O. Gonzalez. 2008. Impact of depression and anxiety on quality of life, health behaviors, and asthma control among adults in the United States with asthma, 2006. Journal of Asthma 45: 123–133.

ten Brinke, A., M.E. Ouwerkerk, A.H. Zwinderman, P. Spinhoven and E.H. Bel. 2001. Psychopathology in patients with severe asthma is associated with increased health care utilization. American Journal of Respiratory and Critical Care Medicine 163: 1093–1096.

The Commonwealth Fund. 2011. Unmet need for mental health care: adults. Performance Snapshots: Tracking Health System Performance. Retrieved February 1, 2011 from: http: //www.commonwealthfund.org/Content/Performance-Snapshots/Unmet-Needs-for-Health-Care/Unmet-Need-for-Mental-Health-Care--Adults.aspx

Tousman, S., H. Zeitz, C. Bristol and L. Taylor. 2007. Development, implementation, and evaluation of a new adult asthma self-management program. Journal of Community Health Nursing 24: 237–51.

World Health Organization (WHO). 2010a. Mental health: strengthening our response. Retrieved February 10, 2011 from: http: //www.who.int/mediacentre/factsheets/fs220/en/index.html

World Health Organization (WHO). 2010b. WHO Strategy for Prevention and Control of Chronic Respiratory Diseases. Geneva, World Health Organization.

Zielinski, T.A. and E.S. Brown. Depression in patients with asthma. pp. 42–50. *In:* E. S. Brown [ed.] 2003. Asthma: Social and Psychological Factors and Psychosomatic Syndromes. Advances in Psychosomatic Medicine. Volume 24. Basel, Karger.

6

Adherence to Guidelines

Guidelines in Acute Severe Asthma

***Hannah K. Bayes*[1,*] and *Neil C. Thomson*[2]**

ABSTRACT

Acute severe asthma represents a common medical emergency accounting for over 75,000 emergency hospital admissions each year in the United Kingdom and two million emergency department visits each year in the United States. Confidential enquires into asthma-related deaths highlight inadequate assessment and management as a factors contributing to adverse outcome. Recognition of the suboptimal management of asthma expedited production of national and international evidence-based guidelines. These guidelines are developed from the scientific literature and the consensus opinion of an expert panel and serve to highlight best practice in acute asthma care. Despite wide dissemination of these guidelines, adherence remains suboptimal. Areas of concern include: inadequate assessment of patients, particularly in measuring PEF and severity level;

[1]Respiratory Medicine, Institute of Infection, Immunity & Inflammation, University of Glasgow, 120 University Place, Glasgow, G12 8TA; Email: hannah.bayes@talk21.com

[2]Respiratory Medicine, Institute of Infection, Immunity & Inflammation, University of Glasgow, Gartnavel General Hospital, 1053 Great Western Road, Glasgow, G12 OYN; Email: neil.thomson@glasgow.ac.uk

*Corresponding author

List of abbreviations after the text.

underutilization of systemic steroids during admission and at discharge; failure to reassess patients and tailor intensity of treatment to severity of attack; and inadequate patient education and arrangement of follow-up at discharge. Poor adherence with the guidelines has predominantly been studied in the emergency department, but is also evident for hospital care and in both pediatric and adult acute asthma care. Poorer outcomes are seen when the guidelines are not adhered to, including increased admission rates, out-patient morbidity and relapse rates. Implementation initiatives, such as Asthma Care Pathways, are now required to enable translation of best-practice guidelines into improved acute asthma care.

INTRODUCTION

Asthma is characterized by chronic airway inflammation and variable airway obstruction that can lead to persistent symptoms as well as intermittent acute exacerbations. Asthma exacerbations (also referred to as acute asthma or asthma attacks) are episodes with progressive onset of increased breathlessness, wheezing, cough, or chest tightness, either alone or in combination. A decline in peak expiratory flow (PEF) occurs during an exacerbation and is a more reliable indicator of the severity of airflow limitation than symptoms alone. Severe exacerbations of asthma are potential life-threatening and represent a common medical emergency. Death is caused by acute respiratory failure due to airflow limitation.

Although asthma mortality has fallen steadily over the last 10-years, over 1,500 people continue to die from asthma each year within the United Kingdom (UK) and annual worldwide deaths are estimated at 250,000. Acute asthma exacerbations remain a common emergency presentation to primary care or emergency departments. Over 75,000 emergency hospital admissions are due to asthma each year in the UK, a quarter of which are in children below 4-years of age. In the United States (US) approximately 2 million emergency department (ED) visits and 500,000 hospitalizations occur each year secondary to acute asthma.

Confidential enquires into asthma-related deaths suggest that 80% of such deaths are preventable and implicate suboptimal assessment and management by patients and medical staff as factors contributing to adverse outcome. Recognition of this suboptimal management expedited the production of evidence-based guidelines, which serve to highlight best practice in acute asthma care.

PRACTICE AND PROCEDURES: GUIDELINES IN ACUTE ASTHMA

Current Acute Asthma Guidelines

Acute asthma guidelines were first developed over two decades ago, based on review of the scientific literature and the consensus opinion of an expert panel. Both national (including the British Thoracic Society (BTS)/Scottish Intercollegiate Guidelines Network (SIGN) and United States (US) National Institute for Health (NIH)) and international (the Global Initiative for Asthma (GINA)) guidelines now exist. Through disseminating information on best practice, they aim to translate research advances into clinical practice, decrease variation in quality of care and, ultimately, bring about improvements in asthma-related morbidity and mortality. Institutes may also develop their own local acute asthma management protocols, often based on the consensus guidelines, to match service provisions.

Within published guidelines, the strength of evidence supporting each recommended intervention is highlighted. The producers of guidelines regularly up-date the recommendations in an evidence-based manner involving systematic review; thus ensuring new research is translated into clinical practice.

Dissemination of Acute Asthma Guidelines

Dissemination of guidelines is a process aimed to inform healthcare providers regarding the existence of guidelines and to ensure knowledge and understanding of their content. Effective dissemination is essential to enable implementation of guideline recommendations. This has been achieved for asthma guidelines via publication in professional journals, specific educational initiatives, inclusion of the guidelines into medical education programs (e.g., advanced life support training), postal distribution of guidelines to relevant healthcare professionals, and provision of easily accessible summary charts which can be used in the acute setting.

Current Guidelines: Assessment of Acute Severe Asthma

Objective assessment of the severity of an exacerbation forms the cornerstone of all acute asthma guidelines (Fig. 1). The grading of severity varies between guidelines, but this is less important than the need for appropriate assessment of the acute asthmatic.

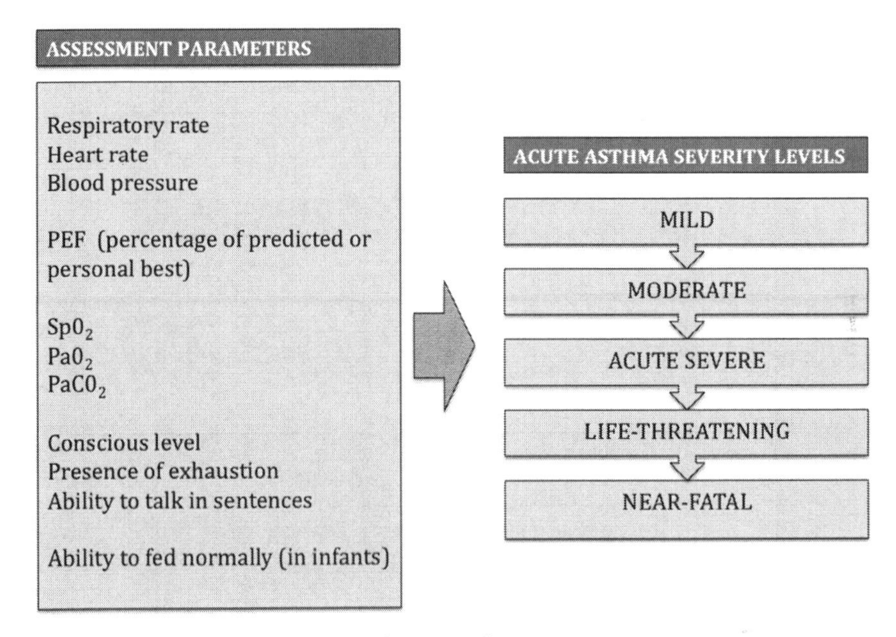

Figure 1 Assessment of severity of acute asthma.
Assessment of the acute asthmatic incorporates a clinical component, measurement of airflow obstruction (most conveniently via PEF rate measurement), oxygen saturations, and, in more severe cases, the assessment of alveolar ventilation via arterial blood gas measurements. Guidelines stipulate criteria based on these parameters to allow classification of the severity of attacks. PEF: peak flow rate. $Pa0_2$: partial pressure of oxygen in arterial blood. $PaC0_2$: partial pressure of carbon dioxide in arterial blood. $Sp0_2$: oxygen saturation measured by pulse oximetry.

Current Guidelines: Treatment of Acute Severe Asthma

Patients with severe or life-threatening exacerbations should be treated in an acute care setting (such as a hospital emergency department (ED)), as they require close monitoring and may require admission. The principles of management of acute asthma are similar in both industrialized and developing countries, including appropriate assessment, oxygen supplementation, regular inhaled bronchodilator treatment, and systemic steroids. The aims of treatment are to promptly relieve bronchocontriction and correct hypoxemia, as well as to plan the prevention of future exacerbations. The intensity of treatment is guided by the severity of the attack and response to treatment (Fig. 2). Treatment algorithms exist for both adult and pediatric patients. There exists some differences between guidelines, for example the NIH guidelines recommend the use of helium-oxygen (heliox)-driven nebulization in patients with impending respiratory failure; this is not contained within other guidelines.

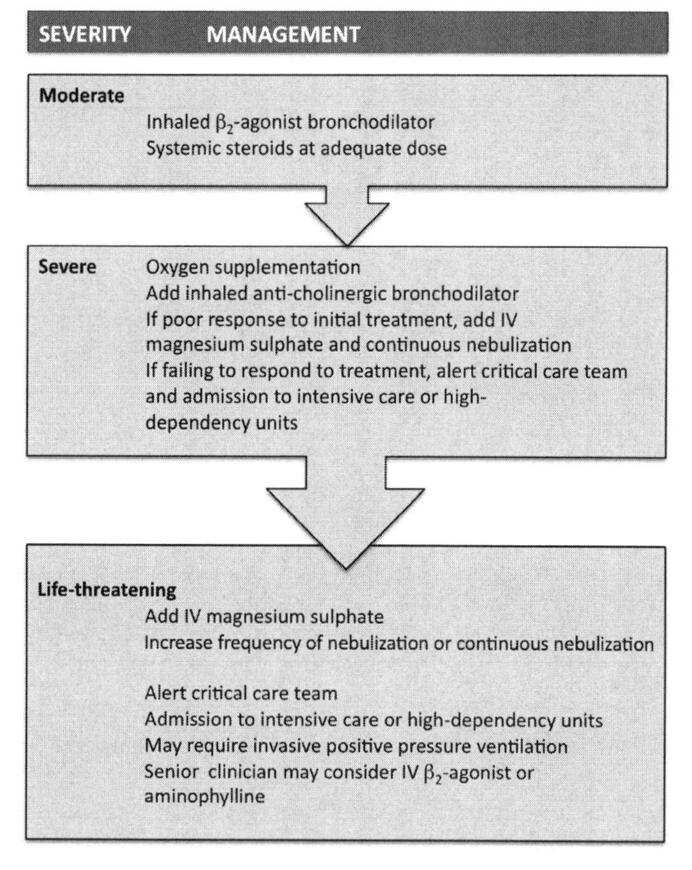

Figure 2 Principle treatments guided by attack severity assessment in adults. Interventions in acute asthma are guided by attack severity and those with severe or life-threatening asthma should be managed in an acute care setting. IV: intravenous.

Current Guidelines: Appropriate Discharge of Patients with Acute Asthma

Guidelines suggest criteria for safe discharge of patients from the ED and depend on severity, response to treatment and presence of any adverse psychosocial factors. Education prior to discharge from either the ED or in-patient admission is essential component of management to prevent further attacks. It is recommended that medical follow-up be arranged at discharge, including in primary care and specialist respiratory review.

ADHERENCE TO GUIDELINES IN ACUTE SEVERE ASTHMA

Initial hopes were that development of acute asthma guidelines would result in improved clinical practice. Adherence to best practice guidelines is frequently used as a benchmark for quality-of-care and, in the case of acute asthma, may encompass a variety of components (Table 1). However there exists continuing concern that the adherence to guidelines, and thus care of acute severe asthma, remains suboptimal. Gaps between management guidelines and practice have been documented worldwide.

Referral from Primary Care

As current guidelines recommend that the majority of patients with acute severe asthma are managed in an acute care setting, the literature predominately assesses the appropriateness of management in secondary care. General practitioners however play a key role in the initial assessment, management and referral of patients, particularly given that 90% of acute asthma is initially managed in primary care and the majority of asthma-related deaths occur prior to hospitalization.

In the UK during the 1990's, studies of general practitioners' management of acute asthma demonstrated underuse of nebulized bronchodilators and systemic steroids. Of greater concern was the management of life-threatening attacks, with only 71% of patients

Table 1 Components of acute severe asthma care to which adherence can be assessed.

Fundamental components of acute severe asthma care
Primary Care
• Appropriate assessment
• Recognition of attack severity
• Immediate treatment
• Referral to acute care setting
Emergency Department (ED)
• Assessment of attack severity
- objective measurement of lung function (PEF or spirometry)
- measuring vital signs required to assess severity
- accurate classification of attack severity
• Treatment in the ED
- treatment in concordance with attack severity
- reassessment after initial treatment
• Appropriate discharge vs. hospitalization
Medical Wards/Hospital Care
• Monitoring, including PEF rate measurement
• Appropriate treatment according to attack severity
Asthma Education & Relapse Prevention
• Inhaler technique checked
• Provision of PEF meter
• Provision/review of Asthma Action Plan
• Review of factors precipitating exacerbation
Discharge Medication
• Course of systemic steroids
• Initiation or continuation of inhaled steroid
• Continuation of bronchodilator
Follow-up arranged at discharge
• Prompt primary care review
• Specialist respiratory review
• Alert primary care practice to recent discharge

Asthma management guidelines include a range of evidence-based recommendations to which adherence, and thus quality of care, can be assessed. PEF: peak expiratory flow.

receiving emergency bronchodilation and only 21% of patients in this category being referred to hospital (Pinnock et al. 1999).

Poor understanding of the acute asthma guidelines appears to be an ongoing problem in primary care; knowledge of the UK BTS/SIGN acute asthma guidelines remains limited, with worryingly only 39% of healthcare professionals able to identify signs of life-threatening asthma and similar results for both general practitioners and primary care nurses (Pinnock et al. 2010). The recording of PEF rate in acute asthmatics attending primary care has also continued to be reported as suboptimal.

Recording of Clinical Parameters in Acute Asthma

Patients with a severe or life-threatening asthma exacerbation may not be distressed and may not have all the severity criteria delineated in the guidelines. The presence of any one abnormality within a given severity level should indicate the severity of the attack, and thus systematic assessment is paramount. A significant number of observational studies have examined the recording of clinical parameters indicative of exacerbation severity in a variety of secondary care settings. In general the parameters of respiratory rate, pulse rate and oxygen saturations are well documented in patients presenting acutely. Arguably this may simply reflect the 'routine' observations taken on patients attending an acute care setting. More concerning is the consistent underuse of PEF rate assessment, with an assessment of PEF rate generally performed in only 50–60% of adult patients. Children over 5-years of age are likely to be able to cooperate with PEF recording, yet a large UK study of hospital admissions for childhood asthma demonstrated that this was recorded in only 36% of patients (Hilliard et al. 2000), and a Canadian ED study also found PEF documentation in only 27% of pediatric participants (Lougheed et al. 2009).

Assessment of Attack Severity

Definitive classification of asthma attack severity is fundamental to all current guideline recommendations. Not only should a severity assessment be made, but it should also be accurate. Unfortunately a formal statement of severity is rarely recorded in patients attending the ED or admitted to hospital. Our study of acute asthma admissions to a UK city teaching hospital, demonstrated recording of the clinical parameters required to make a severity assessment was high (Fig. 3) (Bayes et al. 2010). However, these parameters were not translated into an accurately severity assessment in the majority of cases, with all inappropriate assessments being underestimates of the attack severity and none of the life-threatening asthma attacks (40% of the study cohort) being correctly identified (Fig. 4). Therefore, suggesting that lack of appreciation of attack severity remains a problem even in those patients deemed sufficiently unwell to warrant hospital admission.

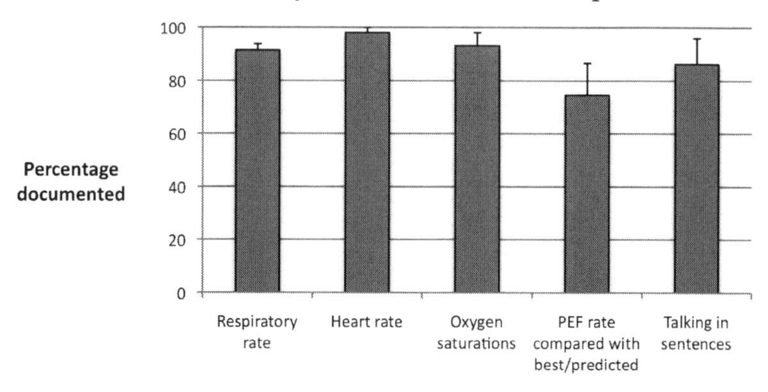

Figure 3 Recording of clinical parameters in adult patients with acute asthma admitted to hospital.
Clinical parameters required to assess the severity of an acute asthma exacerbation were well recorded in admissions to UK teaching hospital. Histogram shows mean performance and standard deviation for two periods of data collection within the hospital. PEF: peak expiratory flow. Data from Bayes et al. 2010.

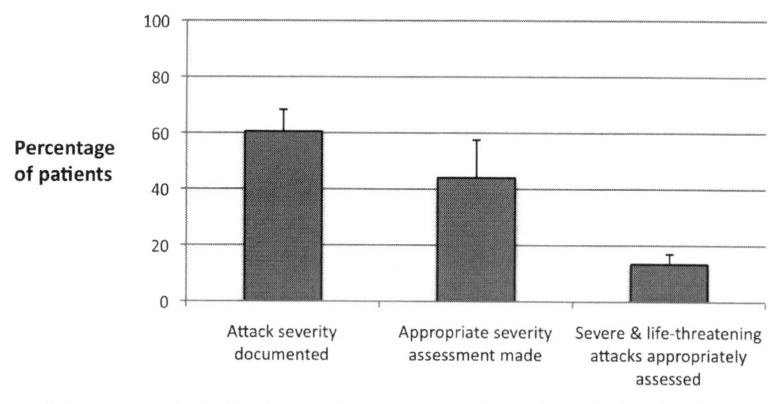

Figure 4 Assessment of attack severity in acute asthmatics admitted to hospital.
Low levels of appropriate attack severity assessment were seen for adult patients with acute asthma admitted to hospital. 'Appropriate severity assessment' denotes percentage of patients who had severity assessment documented and it was consistent with the severity level based on the recorded clinical parameters for that patient. Histogram shows mean performance and standard deviation for two periods of data collection within the hospital. Data from Bayes et al. 2010.

Treatment of Acute Severe Asthma

The strong recommendation for use of bronchodilator therapy in acute asthmatics is generally well adhered to, with the majority of ED audits reporting high levels of inhaled β_2-agonist bronchodilator prescribing. The prescription of anti-cholinergic bronchodilators is more variable; with some centers reporting inappropriate overuse in milder or responding exacerbations, while others report omissions in those with severe and life-threatening asthma.

The prescription of systemic steroids to patients attending the ED is widely reported to be underused. Steroid treatment is recommended in virtually all but the mildest acute asthma exacerbations. Early use results in reductions in mortality, relapses, requirement for hospital admission and need for beta$_2$-agonist therapy. Large studies have reported systemic steroid use in 60–78% of adult patients attending the ED. Pediatric care has also reported underuse, with one in six children attending UK emergency departments not receiving systemic steroids (Hilliard et al. 2000). Although factors such as patient or parent preference, oral steroid taken prior to ED attendance and a milder severity may account for some of these omissions, given the strong evidence for improved outcome with steroid use these levels of underuse appear unacceptable high.

It is important not only to provide appropriate treatments but also to avoid unnecessary investigations and interventions. The majority of infections precipitating an acute asthma exacerbation are viral in etiology. Guidelines therefore emphasis that routine use of antibiotics is not indicated in acute asthma. However one in five acute adult and pediatric patients continue to receive antibiotics unnecessarily when attending the ED in the US (Vanderweil et al. 2008). Current guidelines do not recommend routine chest x-ray (CXR) in acute asthma. However, a recent multi-centre study reported that 30% of pediatric and 43% of adult patients have a CXR performed in the ED (Lougheed et al. 2009). Of concern, a CXR was performed on average as or more often than PEF measurements.

The majority of adherence studies have examined treatment within the ED. However patients with severe attacks failing to improve sufficiently in the ED and life-threatening attacks should be admitted to hospital for ongoing treatment. A national audit in Wales UK, found deficiencies in use of oral steroid and nebulizer therapy in hospital in-patients (Davies et al. 2009). For adult admissions to a UK teaching hospital, we prospectively compared appropriateness of treatment in each attack over the first 24-hours of admission to the BTS/SIGN guidelines. Overall, 60% of patients were inappropriately treated for their attack severity, in 41% this was due to under-treatment (Bayes et al. 2010). Inadequate treatment was most commonly evident in those with life-threatening asthma due to failure to initiate intensive care review, omission of intravenous magnesium or anti-cholinergic bronchodilators.

In acute severe asthma it is important that there is prompt provision of treatment and re-assessment of response to guide further interventions. Emergency department management algorithms in current UK and US guidelines highlight prompt timeframes for delivery of treatments and reassessment, recommendations predominantly based on expert opinion. Studies have largely addressed what treatments are administered, rather than timeliness of provision and reassessment. However, a large US multicenter study of over 4,000 acute asthma presentations to the ED was able to demonstrate that, although appropriate treatments were administered, there were delays in commencing these and, more importantly, low levels of reassessment to response (Fig. 5) (Tsai et al. 2009). This was reflected in a UK audit where only 62% of patients with acute asthma had any form of reassessment in the ED (Davies et al. 2009).

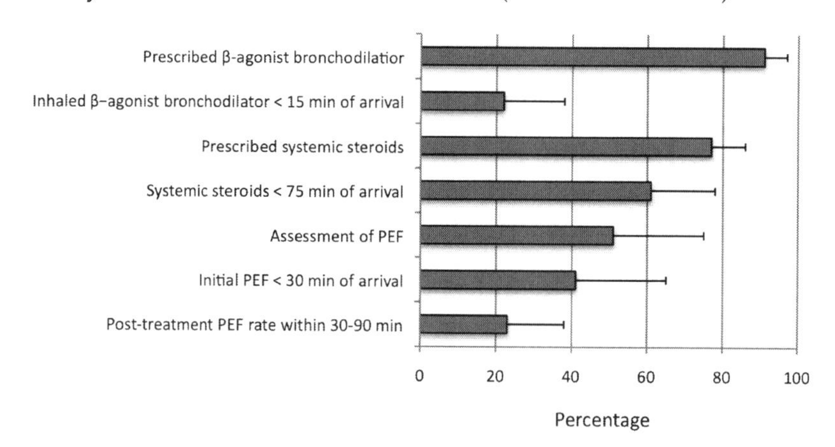

Figure 5 Timeliness of interventions for patients with acute asthma in the emergency department.
Although key treatments for acute asthma may be given in the emergency department, delays in administration and reassessment have been reported. Histogram shows mean performance and standard deviation for 63 US emergency departments. Data from Tsai et al. 2009.

Discharge from the Emergency Department vs. Hospitalization

The minority of ED visits by children and adults with acute asthma result in hospitalization. Asthmatic patients can underestimate the severity of their exacerbation and it is thus the role of the medical team to ensure patients at risk of deterioration are admitted to hospital when appropriate. A large, multicenter, prospective study of ED asthma care in France described high levels of discharge of patients presenting with life-threatening asthma attacks and despite only half of these patients having a PER rate of 50% or greater after initial treatment (Salmeron et al. 2001). Whether these patients represented to the ED or primary care was not described.

At discharge, patients should be prescribed a course of oral steroids, initiated or continued on inhaled steroid and have bronchodilator therapy continued. Studies examining prescribing of oral steroid therapy at discharge from the ED have varied greatly, with steroid courses being provided for 32% to 94% of patients. The majority of data would however suggest underutilization of systemic steroids on ED discharge. Although failure to adhere to guidelines will account for some of this variability, other factors such as patient's own possession of oral steroids, patient or parents preference and milder attack severity will also influence prescription. There is limited evidence examining whether inhaled steroids are continued or introduced for those who may benefit and are discharged from either the ED or hospital. However, in 5,000 pediatric and adult patients discharged from the ED in Canada, only 22% were using an inhaled steroid at discharge (Lougheed et al. 2009). This suggests an under appreciation of guideline recommendations to escalate regular inhaled preventer medication in those presenting with an acute exacerbation.

Education and Follow-up

A proportion of patients with acute asthma will re-attend shortly following discharge from hospital or the ED, with 15% representing within two weeks. However others will remain under-treated, under-monitored and symptomatic without seeking further help.

Providing education, including a written asthma action plan, prior to discharge provides patients with the skills to manage future attacks and has been shown to reduce morbidity and relapses following an exacerbation. The proportion of patients receiving such education has been reported as low in adult and pediatric studies in both the ED and hospital; this includes provision of inhaler device assessment, trigger avoidance, smoking cessation, written asthma action plans and PEF meters/monitoring. Deficiencies in education have previously been found to be more frequent in patients not admitted to hospital.

Appropriate follow-up of patients presenting with an acute asthma exacerbation should be arranged prior to discharge. Failure to refer the majority of these high-risk individuals to either primary or specialist care has been reported in both ED and in-patient studies. For patients admitted to hospital, intended follow-up has been recorded in two-thirds of pediatric admissions with only one-third referred to a consultant-led outpatient clinic (Hilliard et al. 2000) and outpatient department review in 70% of adult admissions (Davies et al. 2009). A recent large Canadian study of over 5,000 ED visits reported low referral rates to specialists, which in total averaged 8% for children and 4% for adults (Lougheed et al. 2009). In the UK, 30% of patients discharged from the ED were provided with early general practice review (Davies et al. 2009).

Variability in Guideline Adherence

Significant variability between institutes in the quality of acute asthma care, judged by guideline adherence, is widely reported. This variability is seen even within geographically close regions. Potential explanations include differences in resources (such as PEF meters/spirometry in the ED or access to specialist asthma care for referral), provision of training for healthcare providers, access to automatic referral systems, and healthcare providers' knowledge of and familiarity with asthma guidelines. Although many of the failings in guideline adherence have been discussed, it is important to recognize that some institutes report good adherence, at least with certain aspects of the guidelines. It is therefore perhaps more unacceptable that such a standard cannot be achieved across all care settings.

Variability in guideline adherence has also been demonstrated between respiratory and non-respiratory hospital specialists. In-patients under the care of a respiratory physician have higher levels of PEF monitoring during admission, greater use of systemic steroids, and are more likely to be discharged with out-patient follow-up arranged (Davies et al. 2009, Bucknall et al. 1988). However, given that acute asthma is a common cause of medical admission and there exists limitations on healthcare resources, many patients with acute asthma will continue to be managed by non-respiratory acute and general physicians.

LIMITATIONS OF STUDIES EXAMINING ADHERENCE

With the exception of a few prospective, multicentre studies, the majority of studies examining adherence to acute asthma guidelines have been retrospective or small prospective studies, often in single centers. The prospective collection of data, for example using admission proformas, means information is more likely to represent what actually occurred in managing the acute asthmatic than if data is collected retrospectively from case notes or observation charts. Generalizability of findings is enhanced by studies which are multicentre in design and large in sample size. However the consistent concordance of studies examining adherence since the asthma guidelines were first produced would strongly suggest that conclusions draw from the current literature are valid.

In addition, there is clearly a significant difference between omission to perform a task and failure to document it. Arguably this may particularly affect assessment of adequate patient education and discharge planning, but also potentially failure to document severity assessment. However documentation in itself is an essential part of management, both to ensure continuity of care and from a medico-legal prospective.

ADHERENCE WITH GUIDELINES AND OUTCOME

Despite consistent documentation of failings in assessment and management in acute severe asthma care, few studies have reported on adverse outcomes such as increased emergency intubations, intensive care admissions or mortality during hospitalization. Presumably, however, care that is in concordance with best practice will result in improved outcomes. Certainly for acute asthma requiring hospital admission, patients who were treated inconsistently with recommended practice reported higher levels of asthma-related morbidity and were ten-times more likely to be re-admitted within the year (Bucknall et al. 1988).

Similar poorer outcome has been reported in the emergency department. Delay in receiving systemic steroids in the ED and failure to prescribe inhaled steroids on discharge is associated with admission to hospital and relapse rates in adults with acute asthma (Lougheed et al. 2009). Whereas patients presenting to the ED who received care which was fully concordant with treatment guidelines were half as likely to require admission (adjusted OR, 0.54; 95% CI 0.41–0.71; P<0.001) (Tsai et al. 2009).

IMPROVING ADHERENCE WITH ACUTE ASTHMA GUIDELINES

It is clear that production and dissemination of guidelines are insufficient to expedite practical improvements in acute asthma care. Successful adoption of guidelines into clinical practice requires implementation initiatives. Comprehensive strategies that involve multiple methods are more likely to be effective in facilitating adherence, including interactive education, audit, feedback and clinical pathways (Fig. 6).

Education alone via didactic lectures and provision of guideline summaries are ineffective in promoting changes in clinical practice. Interactive, practice-based education may be more fruitful. Audit also forms an important component of guideline implementation and in providing specific fed-back to the healthcare team. Indeed acute asthma care has been shown to improve during periods of active audit.

Clinical pathways are an evidence-based approach to guide healthcare professionals in patient assessment and management for a given clinical problem. The aims of clinical pathways are to reduce variation in care, improve resource utilization, and improve quality of care. Asthma clinical pathways (ACP) may include a variety of tools (Fig. 6). These have been developed and studied in both the

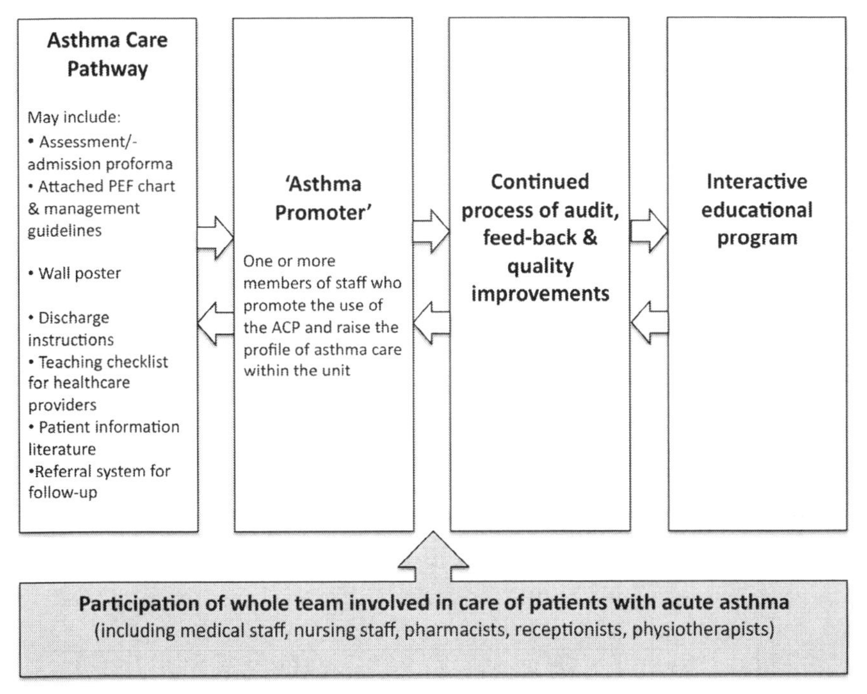

Figure 6 Comprehensive interventions to improve acute asthma guideline adherence.
Improving adherence with acute asthma guidelines is likely to require a multifaceted approach, involving the whole care team. PEF: peak expiratory flow.

ED and in-patient settings. Evidence for their use within the ED is strong. A recent Canadian ED multicenter study of an adult ACP demonstrated significant improvements in assessment, prescription of bronchodilators, administration of systemic steroids both in the ED and on discharge, education and follow-up (Lougheed et al. 2009). The use of the ACP and improved outcomes were achieved with a minimal increase in stay of 16-min within the ED, thus suggesting their use is practical within the time-constraints of current acute care provision. Clinical pathways have also been used for in-patient pediatric asthma management, where they appear effective in reducing length of stay and hospital costs. Evidence as to whether they also result in improved clinical outcomes, such as asthma education and reduce readmission rates, is as yet lacking.

Guideline committees and asthma organizations are increasingly recognizing the need to focus efforts on implementation strategies to overcome the gaps between best practice guidelines and clinical care. Resources for heath professionals are now readily available, including advice on asthma training programs, developing a treatment pathway, templates for assessment proformas and audit, and patient information literature (http://www.asthma.org.uk/health_professionals/materials_to_help_you_your_patients/emergency_care.html).

KEY FACTS IN ADHERENCE TO ACUTE ASTHMA GUIDELINES

- Asthma is a chronic inflammatory disease of the airways that can cause persistent symptoms as well as acute exacerbations (or acute asthma).
- Acute asthma exacerbations are potentially life-threatening and emergency medical management is required.
- Review of asthma-related deaths demonstrated suboptimal assessment and management to be factors contributing to adverse outcomes.
- Management guidelines have been developed with the aim of improving acute asthma care.
- National and international guidelines are available and recommend treatments based on reviewing the best available research.
- Despite wide dissemination, there exists evidence of suboptimal adherence to asthma guidelines in the emergency department and hospitals.
- Poor adherence is seen for the assessment of patients, prescription of treatments, patient education to prevent relapses and in organizing follow-up.
- Poor adherence with the guidelines can result in poorer outcome for the patient.
- Strategies are required to ensure implementation of acute asthma guidelines.

SUMMARY POINTS

- Acute severe asthma is a common medical emergency with inadequate assessment and management associated with asthma-related deaths.

- National and international evidence-based guidelines exist to highlight best-practice in acute asthma care.
- Despite wide dissemination of acute asthma guidelines, there exists long-standing suboptimal adherence to the recommendations.
- Suboptimal adherence has been studied and reported most frequently in emergency department care, but is also evident in patients admitted to hospital and in both pediatric and adult populations.
- Areas of concern include: inadequate assessment of patients, particularly in measuring PEF and severity level; underutilization of systemic steroids during admission and at discharge; failure to reassess patients and tailor intensity of treatment to severity of attack; and inadequate patient education and follow-up at discharge.
- Poorer outcomes are seen when the guidelines are not adhered to, including increased admission rates, outpatient morbidity and relapse rates.
- Implementation initiatives, such as Asthma Care Pathways, are now required to enable translation of best-practice guidelines into improved care.

DEFINITIONS

Audit: a process of improving patient care and outcome that involves reviewing management against defined criteria and implementing changes.

Alveolar ventilation: the amount of air that reaches the alveoli and is available for gas exchange.

Asthma action plan: a set of individualized instructions that details how a person with asthma should manage his or her asthma at home.

Bronchoconstriction: the process by which bronchi (passages that conduct air in the lungs) are constricted or narrowed and thus decrease airflow in the lungs.

Bronchodilators: a substance that dilates the bronchi and increases airflow in the lungs.

Hypoxemia: insufficient oxygenation of arterial blood.

Heliox: a gas composed of a mixture of helium and oxygen, which causes less airway resistance than air and therefore reduces the work of breathing.

Observational study: a type of study in which individuals are observed or certain outcomes are measured, but the investigator does nothing to affect the outcome.

Peak Expiratory Flow (PEF): the maximum flow generated during expiration performed with maximal force and started from a full inspiration. It measures airflow through the bronchi and therefore the degree of airways obstruction. PEF rates are compared with the patient's best previous measurement or, if unknown, their predicted PEF based on sex, age and height.

Systematic review: a review of the literature relevant to a particular research question, which aims to identify, appraise, synthesis and summarize the findings of all available relevant evidence.

Spirometry: a means of assessing lung function by measuring the volume of air inhaled and exhaled by the lungs using a machine called a spirometer.

LIST OF ABBREVIATIONS

ACP	:	asthma clinical pathway
BTS	:	British Thoracic Society
CI	:	confidence interval
CXR	:	chest x-ray
ED	:	emergency department
GINA	:	Global Initiative for Asthma
NIH	:	National Institute for Health
OR	:	odds ratio
$PaC0_2$	:	partial pressure of carbon dioxide in arterial blood
$Pa0_2$	:	partial pressure of oxygen in arterial blood
PEF	:	peak expiratory flow
SIGN	:	Scottish Intercollegiate Guidelines Network
$Sp0_2$	:	oxygen saturation measured by pulse oximetry
UK	:	United Kingdom
US	:	United States

REFERENCES

British Thoracic Society/Scottish Intercollegiate Guidelines Network (2008 (Updated 2009)). "British Guideline on the Management of Asthma." Retrieved January, 2011, from http://www.brit-thoracic.org.uk/clinical-information/asthma/asthma-guidelines.aspx.

Bayes, H.K., O. Oyeniran, M. Shepherd and M. Walters. 2010. "Clinical audit: management of acute severe asthma in west Glasgow." Scott Med J 55: 6–9.

Bucknall, C.E., C. Robertson, et al. 1988. "Differences in hospital asthma management." Lancet 1: 748–750.

Davies, B.H., P. Symonds, R.H. Mankragod and K. Morris. 2009. "A national audit of the secondary care of "acute" asthma in Wales—February 2006." Respir Med 103: 827–838.

Global Initiative for Asthma (Updated 2009). "GINA Report, Global Strategy for Asthma Management and Prevention." Retrieved January, 2011, from http://www.ginasthma.com

Hilliard, T.N., H. Witten, I.A. Male, S.L. Hewer and P.C. Seddon. 2000. "Management of acute childhood asthma: a prospective multicentre study." Eur Resp J 15: 1102–1105.

Lougheed, M.D., N. Garvey, K.R. Chapman, L. Cicutto, R. Dales, A.G. Day, W.M. Hopman, M. Lam, M.R. Sears, K. Szpiro, T. To and N.A. Paterson. 2009. "Variations and gaps in management of acute asthma in Ontario emergency departments." Chest 135: 724–736.

Lougheed, M.D., J. Olajos-Clow, K. Szpiro, P. Moyse, B. Julien, M. Wang and A.G. Day. 2009. "Multicentre evaluation of an emergency department asthma care pathway for adults." CJEM 11: 215–229.

National Heart, Lung, and Blood Institute, National Institutes for Health. (2007). "Expert Panel Report (EPR 3); guidelines for the diagnosis and management of asthma." Retrieved January, 2011, from http://www.nhlbi.nih.gov/guidelines/asthma/

Pinnock, H., S. Holmes, M.L. Levy, R. McArthur, and I. Small. 2010. "Knowledge of asthma guidelines: results of a UK General Practice Airways Group (GPIAG) web-based 'Test your Knowledge' quiz." Prim Care Respir J 19: 180–184.

Pinnock, H., A. Johnson, P. Young and N. Martin. 1999. "Are doctors still failing to assess and treat asthma attacks? An audit of the management of acute attacks in a health district." Respir Med 93: 397–401.

Salmeron, S., R. Liard, D. Elkharrat, J. Muir, F. Neukirch and A. Ellrodt. 2001. "Asthma severity and adequacy of management in accident and emergency departments in France: a prospective study." Lancet 358: 629–635.

Tsai, C.L., A.F. Sullivan, J.A. Gordon, R. Kaushal, D.J. Magid, D. Blumenthal and C.A., Camargo Jr. 2009. "Quality of care for acute asthma in 63 US emergency departments." J Allergy Clin Immunol 123: 354–361.

Vanderweil, S.G., C.L. Tsai, A.J. Pelletier, J.A. Espinola, A.F. Sullivan, D. Blumenthal and C.A., Camargo Jr. 2008. "Inappropriate use of antibiotics for acute asthma in United States emergency departments." Acad Emerg Med 15: 736–743.

7

Inhalation Techniques Associated with Different Devices Used in the Treatment of Asthma

Masaya Takemura

ABSTRACT

Inhaled drug delivery is the cornerstone of asthma management. In the past, attention has focused on the development of appropriate formulations and inhalation devices to optimize delivery of asthma drugs, resulting in a wide range of available inhalers. This proliferation of inhaler devices gives prescribers many choices in terms of asthma medications; however, different inhalers may be confusing to patients. Currently, physicians have a choice of 3 types of inhaler devices: pressurized metered-dose inhalers (pMDIs), dry powder inhalers (DPIs), and nebulizers. A number of evidence-based guidelines have concluded that there is no difference in the delivery of treatments from different inhaler devices when used correctly. However, many patients do not follow the proper inhalation technique, which can be critical to the efficacy of treatment. Each inhaler device has advantages and disadvantages, and no single inhaler device can satisfy all patients' needs. For example, pMDIs are portable and compact, but require coordination of breathing and device actuation, which some patients may find difficult. DPIs are breath-actuated and no patient coordination

Respiratory Disease Center, Tazuke Kofukai Medical Research Institute, Kitano-Hospital, 2-4-20 Ohgimachi, Kita-ku, Osaka, 530-8030, Japan;
Email: m-takemura@kitano-hp.or.jp

List of abbreviations after the text.

is required, but they are not suitable for patients with insufficient inspiratory flow. Nebulizers require minimal patient cooperation and coordination and may be useful for acute asthma exacerbations, but these devices are cumbersome and time-consuming to use and maintain. Patient education regarding the proper inhalation technique is the most critical factor when selecting the most suitable device for each patient. This chapter reviews the proper use of inhalers and addresses several approaches for improving and maintaining adherence to proper inhaler use.

INTRODUCTION

Considering the large surface area of the lung and the absence of first-pass metabolism, the alveolar and bronchial epithelium is a unique site for drug absorption. Inhalation therapy is now widely recognized as the best way of administering drugs for deposit into the lungs of patients with asthma. In the last 50 years, there has been a focus on developing appropriate formulations and inhalation devices to optimize the delivery of asthma medications, particularly corticosteroids and β_2 agonists, into the airways. Currently, more than 100 different inhaler/drug combinations are available for the treatment of asthma. The prevalence of these inhalers has resulted in a wide range of choices for the healthcare provider and in confusion for both clinicians and patients regarding how to use these devices correctly. The efficacy of these drugs depends on their proper inhalation into the lungs. A poor inhaler technique can markedly reduce the proportion of drug that reaches the lung. Several studies have shown that incorrect inhalation technique is associated with poor asthma outcomes, not only in terms of suboptimal asthma control, but also an increased risk of emergency department visits and of death due to asthma exacerbations (Giraud and Roche 2002, Hesselink et al. 2001). Poor inhaler technique thus has a substantial impact on both the individual and on society as a whole. The present chapter focuses the inhaler devices, with particular emphasis on inhalation technique, and addresses several approaches for improving and maintaining adherence to proper inhaler use.

INHALER DEVICES

Physicians currently have a choice of 3 types of inhaler devices for lung deposition of drugs: pMDIs, DPIs, and nebulizers. In particular, an increasing variety of pMDIs and DPIs are available. This has been driven in large part by the development of new drug formulations and the ban on chlorofluorocarbon (CFC) propellants for pMDIs. CFC propellants were commonly used propellants in the early days of pMDIs; however, CFC is not environmentally friendly and depletes the ozone layer. In 1989, The Montreal Protocol on Substances that Deplete the Ozone Layer set a timetable for elimination of CFC use, resulting in the development of inhaler devices that did not use CFC propellants. Thus many types of inhalers are now available, but these newly invented devices have different shapes and sizes and require different techniques to load the device prior to inhalation. A recent systematic review of the literature concluded that "when used correctly" there was no difference in clinical effectiveness between these inhaler devices (Dolovich et al. 2005). In practice, however, many patients use the inhaler devices incorrectly. In a large study of pMDI or DPI use, 76% of the patients using a pMDI and 49–55% of those using DPIs made at least one critical error that reduced clinical efficacy (Molimard et al. 2003).

PRESSURISED METERED-DOSE INHALERS (pMDIs)

pMDIs are the most widely prescribed inhaler device, with more than 400 million units produced annually. These devices include the container, the propellant, the formulation (solution in the propellant), the metering valve, and the actuator (Fig. 1). However, these devices have been referred to as the most complex dosage form in medicine. Essential features of correct usage and common errors associated with pMDI use are shown in Tables 1 and 2, respectively. Many studies have confirmed that incorrect use of pMDIs is a widespread problem. More than 30 years ago, Crompton estimated that approximately half of all users were unable to use the pMDI correctly (Crompton 1982), and the current situation appears to be little improved (Giraud and

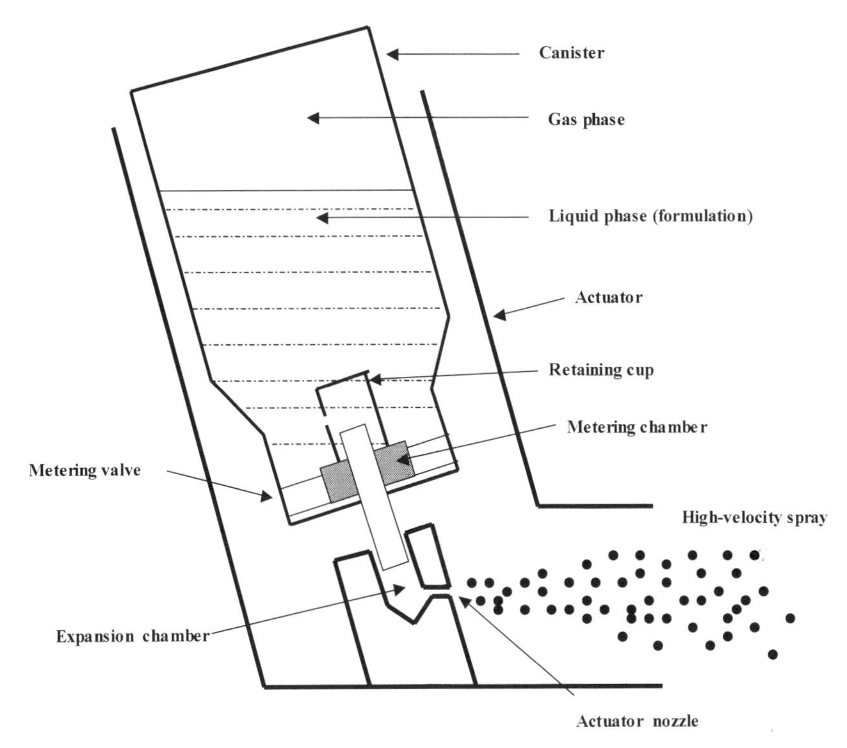

Figure 1 Structure of a pressurized metered-dose inhaler.

Roche 2002). The most common error associated with pMDI use is difficulty coordinating aerosol release (actuation) with inspiration. Poor coordination between actuation and inhalation reduces lung deposition of drugs to 7.2% compared with 22.8% for patients who correctly use the pMDI (Newman et al. 1991). In addition to the importance of coordinating actuation and inspiration, slow and deep inhalation is necessary for better lung deposition (Newman et al. 1995). In a study of asthmatic patients using pMDIs, those who inhaled slowly (30 L/min) and then held their breath for 10 seconds showed the greatest drug deposition in the lung, including both the tracheobronchial and alveolar regions (Newman et al. 1982). However, in general, many patients do not perform a slow inhalation when they use their pMDI (Larsen et al. 1994), and are prone to inhaling too fast (>90 L/min) (Al-Showair et al. 2007). To compensate for these drawbacks in pMDIs, some supplemental devices have been developed and various improvements have been made.

Table 1 Correct technique for using a pMDI*.

1. Take the cap off the inhaler mouthpiece.
2. Shake the inhaler†.
3. Hold the inhaler upright.
4. Breathe out.
5. Place the inhaler mouthpiece between the lips (and the teeth); keep the tounge from obstructing the mouthpiece.
6. Trigger the inhaler while breathing in deeply and slowly (this should be at about 30L/min).
7. Continue to inhale until the lungs are full.
8. Hold the breath while counting 10.
9. Breath out slowly.
*Modified from ref. Broeders et al. 2009, † Unnecessary procedure for HFA-pMDIs.

Table 2 Errors patients make when using pMDIs*.

Error	Percent of patients
1. Failure to co-ordinate actuation and inhalation.	27
2. Inadequate or no breath hold after inhalation.	26
3. Too rapid inspiration/not inhaling forcibly.	19
4. Inadequate shaking/mixing before use.	13
5. Cold Freon effect.	6
6. Actuation at total lung capacity/Not exhaling to residual volume before inhaling.	4
7. Multiple actuations during single inspiration.	3
8. Inhaling trough nose during actuation.	2
9. Exhaling during activation/through the mouth piece.	1
10. Putting wrong end of inhaler in mouth.	<1
11. Holding device in wrong position/incorrectly.	<1
* Modified from ref. Chrystyn and Price 2009.	

a) pMDI with spacer devices

Spacer devices placed between the actuator and the mouth are used to overcome the problems of coordinating actuation and inhalation with the pMDI. Many different commercially available spacer devices are available (Fig. 2), and each has its own characteristics for use. Some spacers are not inhaler-specific, and are designed to allow connection

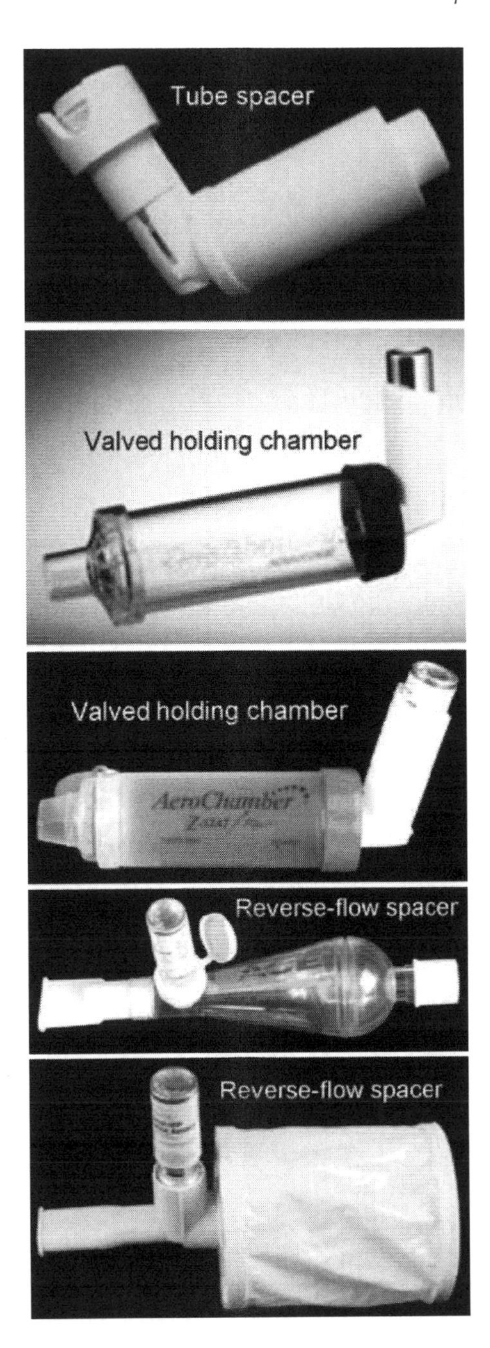

Figure 2 Tube spacer, valved holding chambers, and reverse-flow spacers. Quoted from ref. Hess. 2008 with permission.

with any pMDI. They provide additional space for the aerosol plume to develop, retaining the aerosol until the patient can inhale (Dolovich et al. 2000). They also remove the ballistic component of pMDI use and reduce the deposit of drug into the oropharyngeal area and mouth. Therefore, the Global Initiative for Asthma recommends use of a pMDI with a spacer for the management of asthma (GINA updated 2007). However, several factors may result in inconsistent medication delivery from spacer devices. One of them is the problem of electrostatic charges. Most aerosol particles carry an electric charge that affects aerosol retention within the spacer device (Hinds 1983). To eliminate the electrostatic charge, only one dose should be used per inhalation, and spacers are should be washed in detergent and allowed to dry at least once a week. Coating a spacer with a detergent is a simple method of reducing electrostatic charge, thereby increasing the drug delivery (Chrystyn and Price 2009, Pierart et al. 1999) (Fig. 3). Although using a spacer device is clinically effective, spacers are not easily portable (e.g., they do not fit into the pants or shirt pocket) and are least preferred by patients.

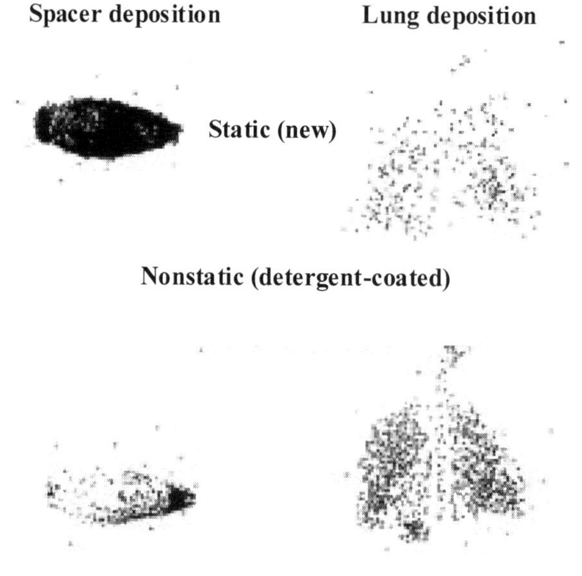

Figure 3 Typical deposition patterns of radioaerosol in a new, untreated valved holding chamber (VHC) (upper left) and a subject's lung after inhalation from an untreated VHC (upper right), and with a detergent-coated VHC (lower left) and the same subject's lungs after inhalation from a treated VHC (lower right). (Reproduced from ref. Pierant F et al. with permission).

b) Breath-actuated pMDIs

A further advancement in pMDIs technology came in the form of a breath-activated pMDIs launched in the United Kingdom in 1989. Breath-actuated pMDIs also compensate for problems of coordinating actuation and inhalation. They have a conventional pressurized canister and are equipped with a flow-triggered system driven by a spring that releases the dose during inhalation. Price et al. reported that patients using breath-actuated pMDIs were prescribed 25% less short-acting β_2-agonists and used up to 64% less oral steroids and up to 44% less antibiotics compared with patients using conventional pMDIs (Price et al. 2003). However, breath-actuated MDIs have the disadvantage of requiring a relatively higher inspiratory flow of approximately 30 L/min to trigger drug release (Fergusson et al. 1991).

c) pMDI with ultrafine particles

When CFC propellant gases were banned, pharmaceutical companies replaced CFCs with hydrofluoroallanes (HFA) in pMDIs. Interestingly, use of an HFA propellant overcomes several of the issues with CFC-MDIs. The formulation in the HFA-MDIs has a warmer temperature, overcoming the cold Freon effect, and is in a solution with absolute ethanol, resulting in emission of ultrafine particles and the lack of need for shaking before use. Combined with an improved actuator nozzle, beclomethasone dipropionate (BDP) with HFA (Qvar® 3M Drug Delivery Systems, Northridge, CA) and ciclesonide (CIC) with HFA (Alvesco® Teijin Pharma, Tokyo, Japan) produce an extra fine aerosol that has a mass median aerodynamic diameter around 1.1 μm. This is in contrast the particle size of 3.5 μm with CFC-MDIs. The proportion of the dose delivered to the lung from an HFA-pMDI is also higher (50–60%) than with a CFC-MDI (less than 15%) (Leach et al. 1998). In addition, because lung deposition of the drug is high, less drug is deposited in the mouth and throat (Leach et al. 1998)

DRY POWDER INHALERS

DPIs have been available since 1970s and were developed to make inhalation easier compared with pMDIs. These devices eliminate the need to coordinate inhalation and actuation (Newman and Busse 2002, Crompton 1982). As opposed to the maneuvers required with a pMDI, one of the most important factors for the proper use of a DPI is deep and forceful inhalation through the device, resulting in better disaggregation, finer particles, and consequently greater transfer of respirable drug particles to the airways (Borgstrom et al. 1994). With DPIs, the respirable particle fraction and consequent drug deposition are dependent on the inspiratory flow rate achieved by the patient. This dependency has been clearly shown with the Turbuhaler® (AstraZeneca, Lund, Sweden). If patients inhale maximally using the Turbuhaler® at the start of the inhalation maneuver, most of the particles produced are 1 and 6 μm diameter and would be deposited as such within the lungs. However, in cases in which inhalation starts slowly and gradually increases, the diameter of the emitted particles increases substantially, resulting in greater drug deposition in the mouth and oropharynx (Everard et al. 1997). The requirement for an initially high inspiratory flow differs in every device (Chrystyn 2003). Table 3 summarizes the critical steps for using 7 currently available inhalers.

There are a number of DPIs available and these devices tend to be complex, with multiple steps required for dose administration and differences between device designs. Molimard et al. examined the specific misuse of each DPI device (Molimard et al. 2003). For example, patients using the Aerolizer® sometimes forgot to insert a capsule, did not press and release both buttons, and exhaled away from the mouthpiece. Errors associated with the Diskus® included failure by patients to hold the mouthpiece in the correct direction, not sliding the lever back as far as possible, not exhaling away from the mouthpiece, and not inhaling through the mouthpiece. With the Turbuhaler®, many patients failed to hold the inhaler in the upright position for grip rotation, did not rotate the grip first clockwise and then counter-clockwise until a "click" was heard, and failed to exhale away from the mouthpiece.

Table 3 Critical steps for using dry powder inhalers*.

Rotahaler®	Diskhaler®	Diskus® (Accuhaler)	Turbuhaler®	Aerolizer®	Twisthaler®	Handihaler®
1. Insert capsule	1.Remove mouthpiece cover	1. Open the device	1. Twist and remove cover	1. Remove cover and twist to open inhaler	1. Keep inhaler straight up when removing cap	1. Remove mouthpiece cover
2. Twist device to break capsule	2. Pull tray out from device	2. Slide the lever	2. Hold inhaler upright (mouthpiece up)	2. Peel back blister and take out capsule	2. Twist cap counterclockwise to lift off cap	2. Take capsule from package
3. Keep device level while rapidly inhaling dose	3. Place disk on wheel (numbers up)	3. Keep the device level while inhaling dose	3. Turn grip right, then left, until it clicks	3. Place capsule in the chamber in the inhaler	3. Exhale fully away from inhaler	3. Place capsule into the inhaler
4. Breath-hold	4. Rotate disk by sliding tray out and in	4. Exhale away from device to residual volume	4. Exhale away from device to residual volume	Twist the inhaler closed	4. Inhale rapidly and fully	4. Close the inhaler
5. Remove device from mouth and exhale away from device	5. Lift back of lid until fully upright so that needle pierces both sides of blister	5. Inhale rapidly and fully	5. Inhale rapidly and fully. Inhaler may be held upright or holizontal for this step	5. Press blue buttons on both sides to pierce capsule	5. Remove from mouth	5. Press button so that needle pierces both sides of capsule
6. Store device in a cool place	6. Keep device level while rapidly inhaling dose	6. Breath-hold	6. Breath-hold	6. Fully exhale away from device	6. Breath-hold	6. Keep device level while inhaling dose rapidly and fully
	7. Breath-hold	7. Remove device from mouth and exhale away from device	7. Remove device from mouth and exhale away from device	7. Tilt head slightly back, inhale rapidly and fully	7. Wipe mouthpiece dry	7. Breath-hold
	8. Remove device from mouth and exhale away from device	8. Store device in a cool dry place	8. Replace cover and twist to close	8. Breath-hold	8. Arrow in line with dose counter	8. Remove device from mouth and exhale away from device
	9. Brush off any powder remaining wiyhin device once every week		9. Store device in a cool dry place	9. Twist open inhaler and dispose of capsule	9. Twist cap clockwise until you hear click	9. Brush off any powder remaining whithin device once every week
	10. Store device in a cool dry place			10. Store device in a cool dry place	10. Store device in a cool dry place	10. Store device in a cool dry place

* Quoted and modified ref. Fink and Rubin 2005

NEBULIZERS

Nebulizers are the oldest of the currently used aerosol delivery devices. From the 19th century until 1956, compressed-air nebulizers (jet nebulizers) were the only devices that were in common clinical use for administration of inhaled aerosol drugs. Ultrasonic nebulizers, which use high-frequency acoustical energy for aerosolization of a liquid, were introduced in the 1960s. Newer designs of nebulizers equipped with breath-enhanced, breath-actuated devices or having aerosol-storage bags to minimize aerosol loss during exhalation are available on the market. Nebulizers are essentially able to convert any liquid into an aerosol. There is strong evidence that for adults and children with acute asthma, equivalent bronchodilator effects can be obtained using multiple doses from a pMDI with different types of spacers as can be obtained using a nebulized delivery system (Dolovich et al. 2005). However, bronchodilatory therapy via a nebulizer system is often used for the management of asthma exacerbations because it may be regarded as more convenient for medical staff to administer and because less patient education or cooperation is required. Nebulized treatment may be also used for some patients who benefit from very high doses of bronchodilator drugs or for patients who are unable to use other devices due to inability to cooperate (e.g., very young children, infants). However, nebulizer systems are bulky, cumbersome, and time-consuming and require electricity to use.

TRAINING IN THE USE OF INHALERS

The issue of correct use is a critical determinant in maintaining optimal asthma control, as patients who misuse inhalers tend to have less stable asthma than those who use their devices correctly. Therefore, training patients regarding correct inhaler use is very important in clinical practice. Several training devices to optimize patients' breathing when using pMDIs and DPIs have been developed. Among the training aids for correct pMDI use, the 2 Tone-Trainer® (Candy Medical, UK) can improve the pMDI technique by ensuring that a slow inhalation is used. This training aid provides users with audible feedback according to the patients' inhalation rate. It makes a 2-tone sound when patients inhale at >60 L/min, a single tone when they inhale at between 30 and 60 L/min, and no sound when they inhale at <30 L/min. Patients are advised to aim for the single tone and become accustomed to the degree of inspiratory effort needed to achieve this rate. With respect to DPIs, some simple device demonstrators, including Diskus-trainer and Turbuhaler-trainer, are designed to teach correct inhalation flow. If a patient inhales through these devices using the correct inhalation speed, the demonstrator "whistles." The In-Check Dial™ (Clement Clarke International Ltd., UK) is a hand-held inspiratory airflow meter designed to identify the most suitable inhaler device for each individual (Chrystyn 2003). The In-Check Dial™ has a dial top that can accurately simulate the resistance of a variety of inhalers on the market (Chrystyn 2003). The Mag-Flo® (Fyne Dynamics Ltd., UK) is another training device that evaluates a patient's ability to use a variety of DPIs including the Turbuhaler®, the Diskus®, the Handihaler®, and Novolizer® (Meda Pharma Brussels Belgium) devices. When a patient inhales properly, a magnetic flow sensor is activated, which switches on a battery-powered green LED that can be seen by the patient. If the patient inhales too strongly or too slowly, the light goes out. Although these training devices are useful for training patients on how to inhale through a device, they do not teach patients how to hold, prime, and position their inhaler device for optimum benefit.

ACHIEVING LASTING PROPER INHALER TECHNIQUE

In practice, many patients make a number of errors using an inhaler device. However, most patients believe that they are using their inhalers correctly (Giraud and Roche 2002), so the problem is not usually identified unless a healthcare professional asks the patient to physically demonstrate their use of the inhaler. Instruction on inhalation techniques apparently results in more efficient use of inhalers (Giner et al. 2002), but these instructions must be repeated, and the results checked at regular intervals. Patients who receive inhalation instructions at least once more after the initial instruction show a better inhalation technique compared with those given a single inhalation lesson at the time they receive their prescription (De Boeck et al. 1999).

Repeated instruction on correct inhalation use not only affects inhalation technique but also clinical outcomes. Basheti et al.

demonstrated that a simple educational intervention regarding inhalation technique, taking only 2.5 minutes per patient per visit and repeated every 6 months, resulted in improvement of clinical indices, expressed as peak expiratory flow variability and quality of life, as well as in improved inhaler technique in asthmatic patients. However, these outcomes tended to decline over the 3 months during which no further education was given (Basheti et al. 2007). In our previous study, we showed that repeated instruction on inhalation technique contributes to adherence to therapeutic regimens, and found a significant relationship between good adherence to inhalation regimen and better asthma control and quality of life (Takemura et al. 2010). The manner in which instruction is provided is also a key factor that determines the quality of inhalation maneuvers. Several studies have shown that written instructions alone, such as the manufacture's pamphlet, are the least effective and inferior to verbal instruction in teaching the correct inhalation technique, regardless of inhaler type (Nimmo et al. 1993, Corsico et al. 2007, Roberts et al. 1982). Verbal instruction and technique assessment are effective ways to achieve proper inhalation technique (Nimmo et al. 1993, Takemura et al. 2010). Naturally, instruction on proper inhalation technique should be given by medical staff members that are able to use the devices correctly themselves. Medical staff should also be educated on using a variety of inhalers. However, in practice, many medical staff members cannot demonstrate correct use of various inhaler devices. For example, Chopra et al assessed the inhaler technique of registered nurses, respiratory therapists, primary care physicians and pharmacists and revealed average technique scores of 81% for pMDI, 64% for Diskus®, and 50% for Turbuhalers® (Chopra et al. 2002). Whether the expansion of the kind of inhalers is good or not, as a future direction, it may be necessary for all members of the asthma management team to have periodic opportunities to relearn proper inhalation techniques for each inhaler device.

PRACTICE AND PROCEDURES FOR SELECTION OF INHALERS

The choice of the drug delivery device is less clear than the choice of the drug needed. Somewhat surprisingly, medical textbooks and asthma guidelines, while emphasizing the importance of checking patients' inhaler technique, provide little practical guidance regarding the specific instructions that are needed to use different inhalers. Given that there is little difference in clinical efficacy between different device types when used correctly (Dolovich et al. 2005), and no single inhaler device can satisfy the needs of all, it is important to tailor the suitable device to each patient. When selecting an inhaler for an individual patient, it is necessary to consider whether sufficient inspiratory flow or an effective vital capacity maneuver is possible for this patient (Fig. 4). For patients with good coordination but without sufficient inspiratory flow (e.g., <30 L/min), a pMDI may be suitable. For patients with inadequate coordination but with sufficient inspiratory flow (e.g., >30 L/min), a DPI, pMDI + spacer device, or a breath-actuated pMDI are all potential choices. If a patient has

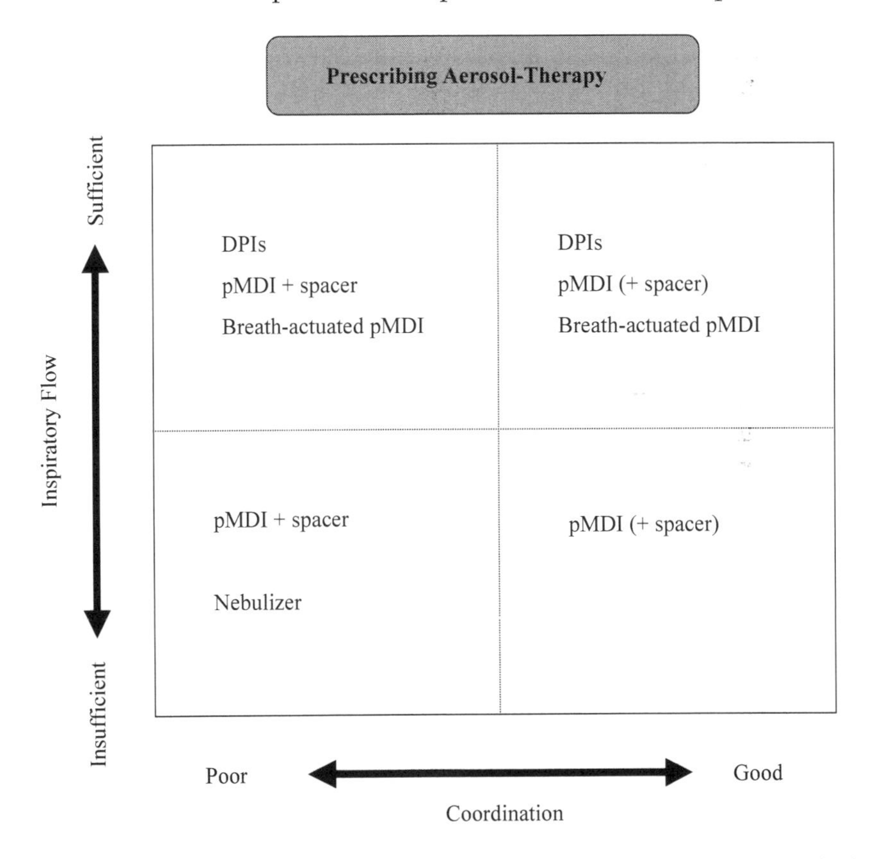

Figure 4 Prerequisites for the selection of an inhalation system in an individual patient.

poor coordination and/or insufficient inspiratory flow, a pMDI with spacer or a nebulizer may be the best choice. Nebulizer use may be debatable for regular treatment as these devices are more expensive, not portable, and time-consuming to use (Broeders et al. 2009); they are thus more appropriate in an acute setting (GINA updated 2007). The ideal prescription for inhaled therapy would use the simplest and most convenient device. Therefore when a patient requires both an inhaled corticosteroid and a β-agonist, a combined inhaler device may be more convenient than multiple actuations from two separate inhalers. If individual combination therapy is unavoidable, combinations of DPIs are preferable to a MDI and a DPI.

KEY FACTS ON INHALATION TECHNIQUES IN ASTHMA

1) Inhalation Techniques in Practice

- A systematic review concluded that when used correctly, there was no difference in clinical effectiveness between pMDIs, DPIs, and nebulizers.
- In a large study of pMDI or DPI use, 76% of the patients using a pMDI and 49–55% of those using DPIs made at least one critical error that reduced clinical efficacy.
- Many medical staff members cannot demonstrate correct use of different inhaler devices. For example, a recent study revealed average technique scores of 76% for pMDI, 64% for Diskus®, and 50% for Turbuhalers®.

2) Optimize the Inhaler for the Patient

- A number of devices are available for training patients in the use of inhalers, such as the 2 Tone Trainer® for pMDIs and the In-Check Dial® for DPIs use. Simple device demonstrators, such as Diskus®-trainer and Turbuhaler®-trainer, are designed to teach correct inhalation flow.
- pMDIs with a spacer device reduce the problem of coordinating actuation and inhalation with pMDIs alone, as well as reduce the amount of drug deposited in the oropharyngeal area. However,

electrostatic charges associated with both aerosols from the inhaler and interior surfaces of the spacer may reduce lung bioavailability.

- The ultrafine particles emitted from HFA-pMDIs are distributed throughout the airway with good penetration, and lung deposition is less affected by any variability in the inhalation flow used by patient.

3) Education for Maintaining Proper Inhalation Technique

- Although asthma education can improve the ability to use inhalers, many patients revert to an incorrect inhalation technique within a short period. Thus, inhalation techniques should be checked regularly.
- Written instruction alone, such as the manufacture's pamphlet, is the least effective method of teaching a patient proper inhalation techniques. Verbal instructions along with actual assessments of the patient's technique are necessary to achieve proper inhalation technique, regardless of inhaler type.

SUMMARY POINTS

- In practice, many patients do not use their inhaler correctly.
- Incorrect inhaler technique is associated with poor asthma control, increased risk of Emergency Department visits, and an increase risk of death due to asthma exacerbation.
- Each inhaler has advantages and disadvantages. No single inhaler device can satisfy the needs of all.
- There is little difference in clinical efficacy between different device types when used correctly.
- Training devices to optimize patients' breathing when using pMDIs and DPIs are available.
- There is no comprehensive guide to assist clinicians in the process of inhaler selection for their patients.
- Important factors to consider when selecting an inhaler device for a patient include the patient's preference, as well as the ability to learn the appropriate inspiratory flow rate.
- Many patients lose the ability to use an inhaler correctly. Inhalation techniques must be checked regularly.

DEFINITIONS AND EXPLANATION OF WORDS AND TERMS

MDI: A metered dose inhaler (MDI) is a device that delivers a specific amount of medication to the lungs, in the form of a short burst of aerosolized medicine that is inhaled by the patient.

Propellants of MDI: The propellant provides the force to generate the aerosol cloud and is also the medium in which the active component is suspended or dissolved.

Cold Freon effect: The sudden impact of cold Freon (aerosol propellants) into the oropharynx can lead to a reflex arrest of inspiration or continuation of inspiration through the nose instead of the mouth.

CFC-pMDIs: In the early days of MDIs, the most commonly used propellants were the chlorofluorocarbons (CFC), including CFC-11, CFC-12, and CFC-114.

Montreal Protocol: The Montreal Protocol on Substances that Deplete the Ozone Layer is an international treaty designed to protect the ozone layer by phasing out the production of numerous substances believed to be responsible for ozone depletion. By 2010, all inhalers containing CFCs were discontinued for hydrofluoroallane (HFA)-pMDIs.

Spacers for MDI: pMDIs are sometimes used with add-on devices referred to as holding chambers or spacers, which are tubes attached to the inhaler that act as a reservoir or holding chamber and reduce the speed at which the aerosol enters the mouth. This makes it easier to use the inhaler and helps ensure that more of the medication gets into the lungs instead of just into the mouth or air.

DPI: A dry powder inhaler (DPI) is a device that delivers medication to the lungs in the form of a dry powder. The patient puts the mouthpiece of the inhaler into the mouth and inhales deeply. Most DPIs rely on the force of the patient's inhalation to get powder from the device and break-up the powder into particles that are small enough to reach the lungs.

Nebulizer: a nebulizer is a device used to administer medication in the form of a mist inhaled into the lungs. The common technical principle for all nebulizers is to use oxygen, compressed air, or ultrasonic power as means of breaking up medical solutions/suspensions into small aerosol droplets, for direct inhalation from the mouthpiece of the device.

GINA: The Global Initiative for Asthma (GINA) is a practical guide of asthma management and prevention launched in 1993. It works with healthcare professionals and public health officials around the world to reduce asthma prevalence, morbidity, and mortality.

LIST OF ABBREVIATIONS

BDP	:	beclomethasone dipropionate
CFC	:	chlorofluorocarbon
CIC	:	ciclesonide
DPI	:	dry powder inhaler
GINA	:	The Global Initiative for Asthma
HFA	:	hydrofluoroalkane
pMDI	:	pressurized metered-dose inhaler
VHC	:	valved holding chamber

REFERENCES

Al-Showair, R.A., S.B. Pearson and H. Chrystyn. 2007. The potential of a 2Tone Trainer to help patients use their metered-dose inhalers. Chest 131: 1776–1782.

Basheti, I.A., H.K. Reddel, C.L. Armour and S.Z. Bosnic-Anticevich. 2007. Improved asthma outcomes with a simple inhaler technique intervention by community pharmacists. J Allergy Clin Immunol 119: 1537–1538.

Borgstrom, L., E. Bondesson, F. Moren, E. Trofast and S.P. Newman. 1994. Lung deposition of budesonide inhaled via Turbuhaler: a comparison with terbutaline sulphate in normal subjects. Eur Respir J 7: 69–73.

Broeders, M.E., J. Sanchis, M.L. Levy, G.K. Crompton and P.N. Dekhuijzen. 2009. The ADMIT series—issues in inhalation therapy. 2. Improving technique and clinical effectiveness. Prim Care Respir J 18: 76–82.

Chopra, N., N. Oprescu, A. Fask and J. Oppenheimer. 2002. Does introduction of new "easy to use" inhalational devices improve medical personnel's knowledge of their proper use? Ann Allergy Asthma Immunol 88: 395–400.

Chrystyn, H. 2003. Is inhalation rate important for a dry powder inhaler? Using the In-Check Dial to identify these rates. Respir Med 97: 181–187.

Chrystyn, H. and D. Price. 2009. Not all asthma inhalers are the same: factors to consider when prescribing an inhaler. Prim Care Respir J 18: 243–249.

Corsico, A.G., L. Cazzoletti, R. de Marco, C. Janson, D. Jarvis, M.C. Zoia, M. Bugiani, S. Accordini, S. Villani, A. Marinoni, D. Gislason, A. Gulsvik, I. Pin, P. Vermeire and I. Cerveri. 2007. Factors affecting adherence to asthma treatment in an international cohort of young and middle-aged adults. Respir Med 101: 1363–1367.

Crompton, G.K. 1982. Problems patients have using pressurized aerosol inhalers. Eur J Respir Dis Suppl 119: 101–104.

De Boeck, K., M. Alifier and G. Warnier. 1999. Is the correct use of a dry powder inhaler (Turbohaler) age dependent? J Allergy Clin Immunol 103: 763–767.

Dolovich, M.A., N.R. MacIntyre, P.J. Anderson, C.A. Camargo, N. Jr. Chew, C.H. Cole, R. Dhand, J.B. Fink, N.J. Gross, D.R. Hess, A.J. Hickey, C.S. Kim, T.B. Martonen, D.J. Pierson, B.K. Rubin and G.C. Smaldone. 2000. Consensus statement: aerosols and delivery devices. American Association for Respiratory Care. Respir Care 45: 589–596.

Dolovich, M.B., R.C. Ahrens, D.R. Hess, P. Anderson, R. Dhand, J.L. Rau, G.C. Smaldone and G. Guyatt. Device selection and outcomes of aerosol therapy: Evidence-based guidelines: American College of Chest Physicians/American College of Asthma, Allergy, and Immunology. Chest 127: 335–371.

Everard, M.L., S.G. Devadason and P.N. Le Souef. 1997. Flow early in the inspiratory maneuver affects the aerosol particle size distribution from a Turbuhaler. Respir Med 91: 624–628.

Fergusson, R.J., J. Lenney, G.J. McHardy and G.K. Crompton. 1991. The use of a new breath-actuated inhaler by patients with severe airflow obstruction. Eur Respir J 4: 172–174.

Fink, J.B. and B.K. Rubin. 2005. Problems with inhaler use: a call for improved clinician and patient education. Respir Care 50: 1360–1374.

Global Initiative for Asthma. Global strategy for asthma management and prevention. http: //www.ginaasthma.com. Updated December 2007.

Giner, J., V. Macian and C. Hernandez. 2002. Multicenter prospective study of respiratory patient education and instruction in the use of inhalers (EDEN study). Arch Bronconeumol 38: 300–305.

Giraud, V. and N. Roche. 2002. Misuse of corticosteroid metered-dose inhaler is associated with decreased asthma stability. Eur Respir J 19: 246–251.

Hess, D.R. 2008. Aerosol delivery devices in the treatment of asthma. Respir Care 53: 699–725.

Hesselink, A.E., B.W. Penninx, H.A. Wijnhoven, D.M. Kriegsman and J.T. van Eijk. 2001. Determinants of an incorrect inhalation technique in patients with asthma or COPD. Scand J Prim Health Care 19: 255–260.

Hinds, W.C. 1982. Aerosol Technology. John Wiley & Sons, New York, USA pp. 284–314.

Larsen, J.S., M. Hahn, B. Ekholm and K.A. Wick. 1994. Evaluation of conventional press-and-breathe metered-dose inhaler technique in 501 patients. J Asthma 31: 193–199.

Leach, C.L., P.J. Davidson and R.J. Boudreau. 1998. Improved airway targeting with the CFC-free HFA-beclomethasone metered-dose inhaler compared with CFC-beclomethasone. Eur Respir J 12: 1346–1353.

Molimard, M., C. Raherison, S. Lignot, F. Depont, A. Abouelfath and N. Moore. 2003. Assessment of handling of inhaler devices in real life: an observational study in 3811 patients in primary care. J Aerosol Med 16: 249–254.

Newman, S.P., D. Pavia, N. Garland and S.W. Clarke. 1982. Effects of various inhalation modes on the deposition of radioactive pressurized aerosols. Eur J Respir Dis Suppl 119: 57–65.

Newman, S.P., A.W. Weisz, N. Talaee and S.W. Clarke. 1991. Improvement of drug delivery with a breath actuated pressurised aerosol for patients with poor inhaler technique. Thorax 46: 712–716.

Newman, S., K. Steed, G. Hooper, A. Kallen and L. Borgstrom. 1995. Comparison of gamma scintigraphy and a pharmacokinetic technique for assessing pulmonary deposition of terbutaline sulphate delivered by pressurized metered dose inhaler. Pharm Res 12: 231–236.

Newman, S.P. and W.W. Busse. 2002. Evolution of dry powder inhaler design, formulation, and performance. Respir Med 96: 293–304.

Nimmo, C.J., D.N. Chen, S.M. Martinusen, T.L. Ustad and D.N. Ostrow. 1993. Assessment of patient acceptance and inhalation technique of a pressurized aerosol inhaler and two breath-actuated devices. Ann Pharmacother 27: 922–927.

Pierart, F., J.H. Wildhaber, I. Vrancken, S.G. Devadason and P.N. Le Souef. 1999. Washing plastic spacers in household detergent reduces electrostatic charge and greatly improves delivery. Eur Respir J 13: 673–678.

Price, D., M. Thomas, G. Mitchell, C. Niziol and R. Featherstone. 2003. Improvement of asthma control with a breath-actuated pressurised metered dose inhaler (BAI): a prescribing claims study of 5556 patients using a traditional pressurised metred dose inhaler (MDI) or a breath-actuated device. Respir Med 97: 12–19.

Roberts, R.J., J.D. Robinson, P.L. Doering, J.J. Dallman and R.A. Steeves. 1982. A comparison of various types of patient instruction in the proper administration of metered inhalers. Drug Intell Clin Pharm 16: 53–55, 59.

Takemura, M., M. Kobayashi, K. Kimura, K. Mitsui, H. Masui, M. Koyama, R. Itotani, M. Ishitoko, S. Suzuki, K. Aihara, M. Matsumoto, T. Oguma, T. Ueda, H. Kagioka and M. Fukui. 2010. Repeated instruction on inhalation technique improves adherence to the therapeutic regimen in asthma. J Asthma 47: 202–208.

8

Treatment of Acute Asthma in the Emergency Setting

Thomas J. Ferro

ABSTRACT

Upon arrival to the emergency department (ED) or other emergent care area, the patient should be evaluated quickly with focused history, physical examination and diagnostic tests to establish a baseline and for triage. Initial treatment in the ED can start during the initial evaluation and should include low-flow supplemental oxygen, intravenous fluid and a short-acting bronchodilator regimen generally delivered via nebulizer. In the unusual event that signs of impending or actual ventilatory failure are present, plans should be made for intubation under controlled conditions; if this is not possible quickly, bi-level positive airway pressure (BiPAP) should be considered as interim therapy if the patient is likely to cooperate with mask therapy, or for patients who decline intubation. Further care should include re-evaluation after initial therapy, a second and a third short-acting bronchodilator treatment at a 20 minute interval if indicated, a minimum of prednisone 40 mg and consideration of additional treatments if indicated. The case findings should be reviewed relative to criteria for admission to determine the optimal disposition. Planning for discharge should include a prednisone taper, an increase in the patient's inhaled corticosteroid (ICS)-based controller regimen, and connection of the patient with their primary care or asthma provider to control the

Formerly of the Department of Internal Medicine, the VCU School of Medicine, Richmond, Virginia 23298; Email: ferrotom@bellsouth.net

List of abbreviations after the text.

patient's asthma, including the avoidance of triggers and adherence to controller or maintenance medications.

Key terms: asthma, betamimetics, emergency department, systemic corticosteroids

INTRODUCTION

Asthma is a common disease in both developed and developing countries, with a prevalence of 1% to 18% in different countries and approximately 300 million affected individuals worldwide (Bateman 2008). It is a frequent cause of emergent visits (office, urgent care, ED). In the USA from 2001 through 2003, asthma accounted for, on an approximate annual basis, 4210 deaths, 504,000 hospitalizations and 1.8 million ED visits (Moorman et al. 2007). The annual rate of ED visits for asthma was 8.8 per 100 persons with current asthma. Rates were higher among children as opposed to adults (11.2 vs. 7.8 visits per 100 persons), among blacks compared to whites (21 vs. 7 visits per 100 persons), and among Hispanics than non-Hispanics (12.4 vs. 8.4 visits per 100 persons). Women made twice the number of ED visits as men (Moorman et al. 2007). Approximately 10% of visits result in hospitalization (Rowe et al. 2009).

Some patients presenting with acute asthma respond rapidly to therapy and can be discharged; others require admission to the hospital for more prolonged observation and treatment (Lazarus 2010). A major goal in the ED is determining which patients can be discharged and which need hospitalization. The recommendations in this article are consistent with the guidelines of the Global Initiative for Asthma (GINA) and the National Asthma Education and Prevention Program Expert Panel Report 3 (NAEPP-3) (Bateman 2008, NHLBI 2007). For further detail, the reader is referred to a recent excellent review article (Lazarus 2010).

CLINICAL EVALUATION OF THE ACUTELY ILL ASTHMA PATIENT

Patients presenting with acute asthma should be triaged quickly based on the severity of the exacerbation and the need for timely intervention (Fig. 1). A focused history and physical examination should be performed. This bedside evaluation should be brief and can be performed while patients receive initial treatment so as not to delay therapy (Lazarus 2010).

The evaluation should focus on findings of acute asthma that are associated with the severity determination (mild-moderate vs. severe) and those that are potentially life-threatening (Table 1).

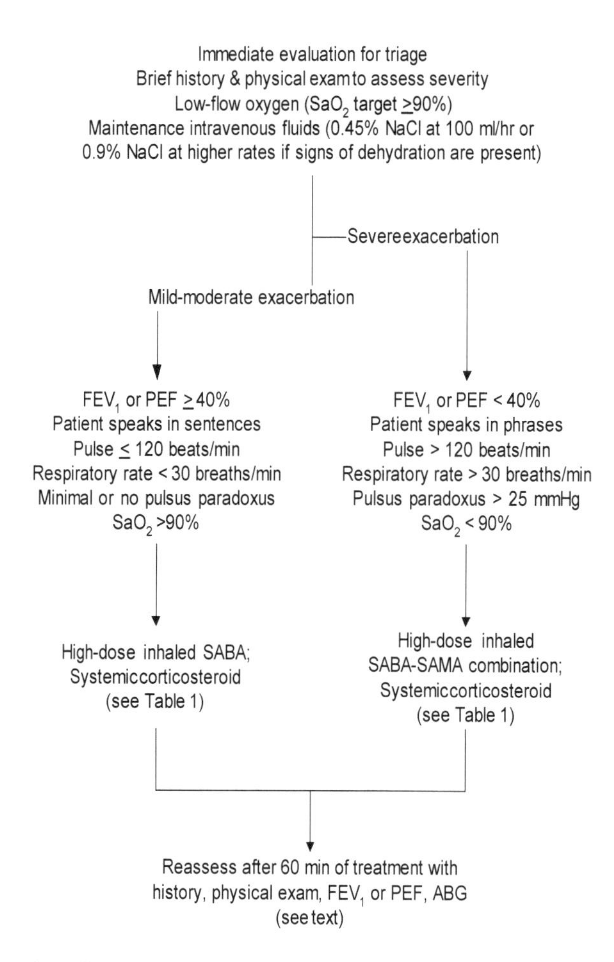

Figure 1 Initial evaluation and treatment of the patient with acute exacerbation of asthma.

Table 1 Findings of Potentially Severe or Life-threatening Acute Asthma that are Potentially Life-threatening.

Finding	Comment
altered mental status	suggestive of hypercapnia
previous intubation or admission to an intensive care unit	
2 or more hospitalizations for asthma during the past year	
3 or more emergency visits in the past year, use of more than 2 canisters per month of SABA	
low socioeconomic status	
various coexisting illnesses	
respiratory distress or patient unable to complete sentences	
pulse greater than 120 beats per min, respirations greater than 30 breaths per min	
pulsus paradoxus	decrease in systolic blood pressure by more than 25 mmHg on inspiration
paradoxical abdominal motion	upward movement of the abdomen on inspiration when supine, suggestive of diaphragmatic fatigue
loud wheezes or the absence of wheezing	absence of wheezing is suggestive of hypoventilation
spirometric forced vital capacity at 1 second (FEV_1) or peak expiratory flow (PEF)	FEV_1 less than 40% of predicted, or PEF less than 40% of personal best, represents severe airflow obstruction, whereas FEV_1 40–70% of predicted, or PEF 40–70% of personal best, may be considered moderate airflow obstruction
arterial blood gas analysis, serum electrolytes, complete blood count	E.g., hypercapnia suggests impending or actual ventilatory failure; hypoxemia is associated with risk for tissue hypoxia and lactic acidosis; elevated serum bicarbonate suggests chronic hypercapnia with chronic ventilatory failure; hyperkalemia might result from acute respiratory acidosis; hypokalemia could result from SABA use; complete blood count might show leukocytosis suggesting infection
chest radiograph and electrocardiogram (ECG)	E.g., chest radiograph could reveal pneumothorax, atelectasis and pneumonia; ECG could also be used as a screen to identify risk for coexisting heart disease

TREATMENT OF ACUTE ASTHMA IN THE ED

Overview and General Measures

All patients should be treated initially as outlined in Fig. 1. If signs of actual or impending ventilatory failure are present, mechanical support should be provided. Aggressive hydration should not be used routinely for acute asthma (NHLBI 2007).

Inhaled Bronchodilators: Short-acting β_2-adrenergic Agonist (SABA) with or without Short-acting Muscarinic Antagonist (SAMA)

Inhaled SABA should be administered as soon as acute asthma is diagnosed. The use of a metered-dose inhaler (MDI) with a valved holding chamber is as effective as the use of a pressurized nebulizer in randomized trials (Dhuper et al. 2008). Most guidelines recommend the use of nebulizers for patients with severe acute exacerbations because of difficulties with inhaler technique; MDIs with holding chambers can be used for patients with mild-to-moderate severity (Lazarus 2010).

A common SABA protocol for emergent use is described in Table 2 (Lazarus 2010). A meta-analysis of results from six randomized trials showed similar effects on both lung function and rate of hospitalization for intermittent vs. continuous administration (Rodrigo and Rodrigo 2002). In contrast, a Cochrane review of eight trials suggested that continuous nebulization resulted in greater improvement in PEF and FEV_1 and a greater reduction in hospital admissions, particularly among patients with severe asthma (Camargo et al. 2003). Levalbuterol, the R-enantiomer of salbutamol (also called albuterol in the US), is effective at half the dose of racemic albuterol/salbutamol, but randomized trials have not consistently shown a clinical advantage of levalbuterol (Qureshi et al. 2005).

The oral or parenteral delivery of SABA is not recommended, because neither has been shown more effective than inhaled SABA, and both are associated with an increase in adverse events (Lazarus 2010). The efficacy and safety in acute asthma of long-acting inhaled β_2-adrenergic agents such as salmeterol and formoterol are unknown.

Table 2 Medications for Treatment of the Patient with Acute Exacerbation of Asthma.

Drug	Regimen	Other Details
SABA		
Salbutamol/albuterol MDI	In emergency department (up to 4 hrs.): 4-8 puffs (90 mcg/puff) every 20 min; thereafter, 2 puffs every 6 hrs as needed.	Patients can use up to 4 puffs every 4 hrs as needed after discharge, but this dose should be used transiently until re-evaluation with possible hospital admission can occur. Adverse effects of SABAs include tachycardia, palpitations, tremor, hypokalemia, paradoxical bronchospasm and others.
Salbutamol/albuterol nebulizer solution	In emergency department (up to 4 hrs.): 2.5–5 mg (supplied as 0.63, 1.25 or 2.5 mg in 3 ml for single use or 5.0 mg in 1 ml to be diluted in normal saline for multiple use) every 20 min over the first hr, then 2.5–10 mg every 1–4 hrs as needed; alternatively: 10–15 mg/hr continuously as needed. Nebulization is not recommended in the outpatient setting, but if preferred on a case-by-case basis, should be used 2.5-5 mg every 6 hrs as needed.	Patients can use 2.5–5 mg every 1 hr as needed, but this dose should be used transiently until re-evaluation with possible hospital admission can occur. Adverse effects of SABAs include tachycardia, palpitations, tremor, hypokalemia, paradoxical bronchospasm and others.
SAMA		
Ipratropium MDI	In emergency department, this should be used only as add-on to SABA therapy (up to 3 hrs.): 2–8 puffs (18 mcg/puff) every 20 min as needed. SAMA is not recommended for use in asthma in the outpatient setting.	Ipratropium is not used as first-line therapy in acute exacerbation of asthma, because of their relatively slow onset of action. Ipratropium has not been shown to provide further benefit as SABA add-on therapy once the patient has been hospitalized. Adverse effects of SAMAs include dryness of the mouth, cough, blurriness of vision and others.

Table 2 contd....

Table 2 contd....

Drug	Regimen	Other Details
Ipratropium nebulizer solution	In emergency department, this should be used only as add-on to SABA therapy (up to 4 hrs.): 0.5 mg (supplied as 0.25 mg/ml, to be diluted in normal saline to volume of 3 ml) every 20 min over the first hr, then as needed; can be used with albuterol/salbutamol in a single nebulization. SAMA is not recommended for use in asthma in the outpatient setting.	Ipratropium is not used as first-line therapy in acute exacerbation of asthma, because of their relatively slow onset of action. Ipratropium has not been shown to provide further benefit as SABA add-on therapy once the patient has been hospitalized. Adverse effects of SAMAs include dryness of the mouth, cough, blurriness of vision and others.
SABA-SAMA Combination		
Ipratropium-salbutamol/ albuterol MDI	In emergency department (up to 3 hrs.): 2-8 puffs (each puff contains 18 mcg ipratropium plus 90 mcg albuterol/salbutamol) every 20 min as needed. SABA-SAMA combination is not recommended for use in asthma in the outpatient setting.	Ipratropium has not been shown to provide further benefit as SABA add-on therapy once the patient has been hospitalized. Adverse effects of SABAs include tachycardia, palpitations, tremor, hypokalemia, paradoxical bronchospasm and others, and those of SAMAs include dryness of the mouth, cough, blurriness of vision and others.
Ipratropium-salbutamol/ albuterol nebulizer solution	In emergency department (up to 4 hrs.): 3 ml (each vial contains 0.5 mg ipratropium and 2.5 mg albuterol in 3 ml) every 20 min for the first hr, then as needed. SABA-SAMA combination is not recommended for use in asthma in the outpatient setting.	Ipratropium has not been shown to provide further benefit as SABA add-on therapy once the patient has been hospitalized. Adverse effects of SABAs include tachycardia, palpitations, tremor, hypokalemia, paradoxical bronchospasm and others, and those of SAMAs include dryness of the mouth, cough, blurriness of vision and others.
Systemic corticosteroid		
Prednisone	40–80 mg/day in one or two divided doses, given for 3–10 days; the dose can be stopped prior to 10 days if the patient is symptom-free or achieves peak expiratory flow 70% of personal best	The dose can be stopped abruptly unless the patient was tapering previously. Adverse effects of short-course systemic corticosteroids include impaired glucose tolerance, hypertension and others.

Whereas SABAs provide dilatory effect to the small airways (<2 mm diameter), anticholinergics such as SAMAs dilate larger airways (Rodrigo and Castro-Rodriguez 2005). The use of the inhaled ipratropium (a SAMA), combined with a albuterol/salbutamol (a SABA), compared to albuterol/salbutamol monotherapy, reduced the hospitalization rate for acute asthma by approximately 25% in patients with severe airflow obstruction; there was no benefit of continuing ipratropium after hospitalization (Plotnick and Ducharme 2000). Based in part on these data, the use of an inhaled combination of ipratropium with albuterol/salbutamol has become common in patients with acute asthma presenting with severe airflow obstruction (Rodrigo and Castro-Rodriguez 2005). A common protocol for the use of SABA-SAMA is described in Table 2 (Lazarus 2010).

Systemic Corticosteroids

Systemic corticosteroids are indicated for acute exacerbation of asthma. The exception may be the patient who has a rapid response to initial SABA therapy. Randomized, controlled trials in acute asthma patients have shown that systemic corticosteroids are associated with a more rapid improvement in lung function, fewer hospitalizations, and a lower rate of relapse after discharge from the ED (Krishnan et al. 2009). Oral prednisone is preferred for patients with normal mental status and without conditions expected to interfere with gastrointestinal absorption, because comparisons of oral prednisone and intravenous corticosteroids have not shown differences in the rate of improvement of lung function or in the length of the hospital stay (Jónsson et al. 1988). A common protocol for the use of prednisone is described in Table 2 (Lazarus 2010).

Ventilatory Support

Patients with signs of actual or impending ventilator failure (e.g., altered mental status, paradoxical abdominal motion and hypercapnia) should be considered for immediate intubation and mechanical ventilation (Fig. 2). Patients intubated and on mechanical ventilation should be transferred to an intensive care unit as quickly as possible.

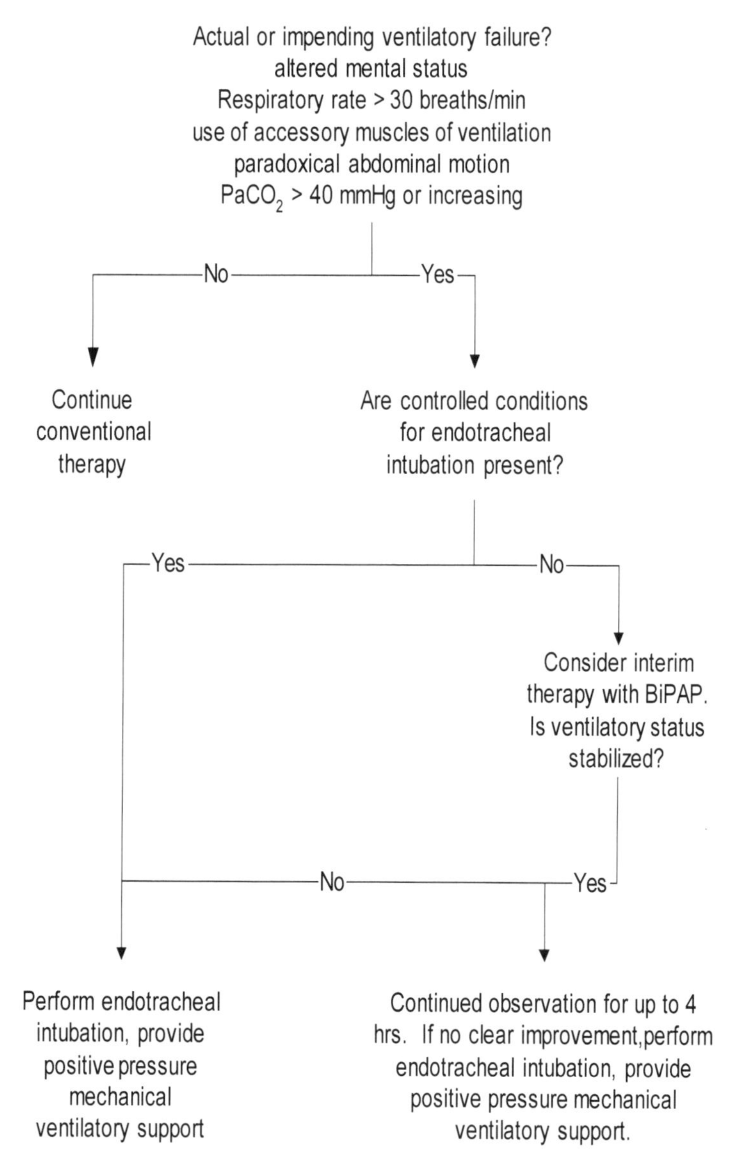

Figure 2 Ventilatory Support for the Patient with Acute Exacerbation of Asthma.

Preliminary studies suggest that noninvasive positive-pressure ventilation may be a consideration for interim therapy in patients likely to cooperate with mask therapy, or for patients who decline intubation. Larger studies are required, however, before this approach can be recommended. In a randomized, sham-controlled trial of bi-level positive airway pressure (BiPAP) in 30 adults with acute exacerbation of asthma, BiPAP was associated with a higher FEV_1 at 4 hours and a lower rate of hospitalization (17.6% vs. 62.5% with sham treatment) (Soroksky et al. 2003). A randomized crossover trial of BiPAP vs. standard care for 2 hours in children with acute asthma showed that BiPAP significantly lowered respiratory rate and improved asthma symptom scores but had no significant effect on other outcomes (Thill et al. 2004).

Other Treatments

Anticholinergic Monotherapy. Inhaled ipratropium is not recommended as a first-line bronchodilator in acute exacerbation of asthma, because of its relatively slow onset of action. Inhaled ipratropium can be added to a SABA, however, as discussed above (Rodrigo and Castro-Rodriguez 2005). A common SAMA "add-on" protocol is shown in Table 2 (Lazarus 2010).

Inhaled Corticosteroids (ICS). ICS cannot substitute for systemic corticosteroids in acute exacerbation of asthma (Edmonds et al. 2003). In contrast, ICS are first-line therapy for long-term asthma control. Therefore, ICS should be prescribed for use once disease is stabilizing. The addition at discharge from the ED of inhaled budesonide to a taper of oral corticosteroids for 5 to 10 days, compared with oral corticosteroids alone, was associated with a 48% reduction in the rate of relapse, and with improvement in the Asthma Quality of Life Questionnaire, at 21 days in a randomized, controlled trial of 1006 consecutively enrolled patients (Rowe et al. 1999).

Antibiotics. Antibiotics should not be used routinely in acute exacerbation of asthma. Acute exacerbation of asthma may be precipitated by bacterial infection (e.g., pneumonia or sinusitis), however, so signs and symptoms of bacterial infection should be given careful consideration and antibiotics should be used if indicated (NHLBI 2007).

Magnesium. A meta-analysis of 24 studies of 1669 patients who received either intravenous magnesium sulfate (15 studies) or nebulized magnesium sulfate (9 studies) showed that intravenous treatment was weakly associated with improved lung function in adults but had no significant effect on hospital admissions; in children, the use of intravenous magnesium sulfate significantly improved lung function and reduced rates of hospital admission; nebulized magnesium sulfate showed minimal effect (Mohammed and Goodacre 2007). The consensus among experts suggests that clinicians consider intravenous magnesium sulfate in patients who have exacerbations of asthma and whose FEV_1 remains less than 40% of the predicted value or PEF remains less than 40% of the personal best (NHLBI 2007, Rowe et al. 2008).

Heliox. Heliox (a mixture of 79% helium and 21% oxygen) has a density that is roughly one third that of air. The low density reduces airflow resistance, especially in regions of the bronchial tree with mostly turbulent flow, presumably reducing the work of breathing and also improving the delivery of inhalational medications. Current guidelines suggest that heliox be considered in patients with severe airflow obstruction who have not had a response to initial treatment, based on a Cochrane analysis of 544 patients in 10 trials (NHLBI 2007, Rodrigo et al. 2006).

Leukotriene Inhibitors. The usefulness of oral leukotriene inhibitors in the acute, emergency setting is unclear. Interest in these agents is based on their ability to increase the FEV_1 within 1 to 2 hours (Liu et al. 1996, Dockhorn et al. 2000). Furthermore, intravenous montelukast significantly improved the FEV_1 at 60 minutes but did not reduce the rate of hospitalization in a randomized, placebo-controlled trial of 583 adults whose FEV_1 remained at 50% or less of the predicted value after 60 minutes of standard care (Camargo et al. 2010).

Methylxanthines. Methylxanthines are no longer considered a standard treatment for acute exacerbation of asthma, because of increased risk of adverse events without improving outcomes (Parameswaran et al. 2000).

Mucolytics. Mucolytic agents are not recommended for acute exacerbations of asthma (NHLBI 2007).

EVALUATION OF THE THERAPEUTIC RESPONSE

Patients should be assessed after the first bronchodilator treatment and again at about 60 minutes, usually after the third SABA treatment

(NHLBI 2007). The assessment should include a survey of symptoms, a search for pertinent physical findings, measurement of either FEV_1 (preferred) or PEF and the measurement of arterial blood gases, at least in the more severe exacerbations (Fig. 3). Approximately two-thirds of patients with acute exacerbation of asthma will meet the

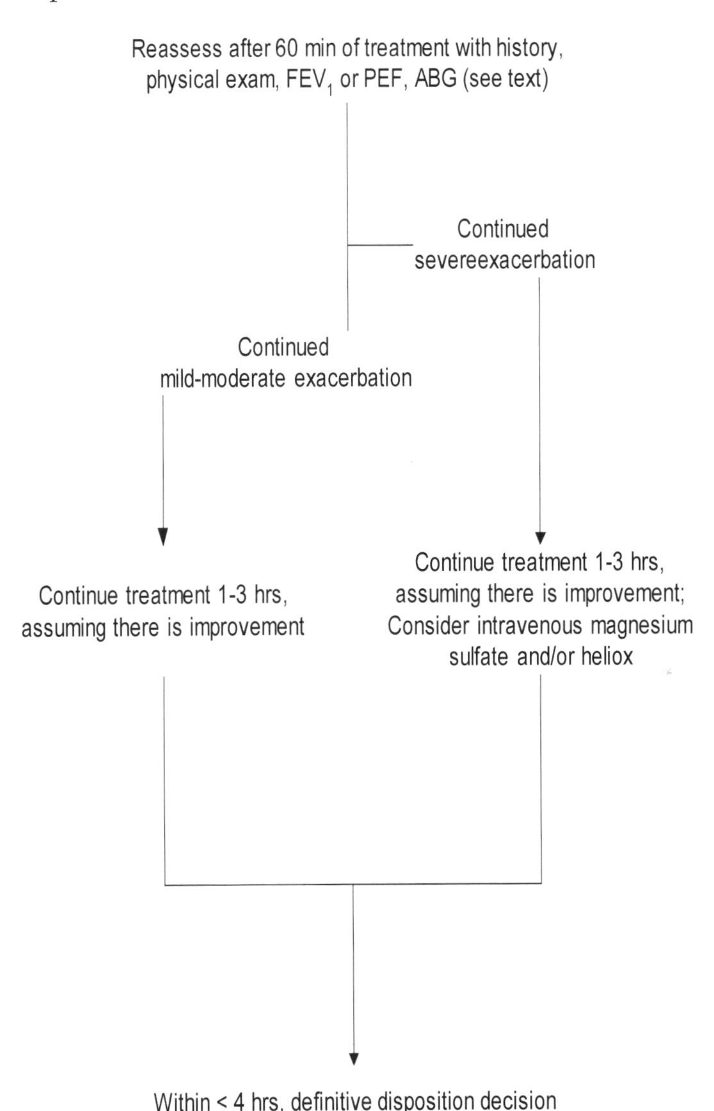

Figure 3 Ongoing Management of Acute Exacerbation of Asthma.

criteria for discharge from the ED (as outlined below) after three doses (Rodrigo and Rodrigo 1998). The degree of improvement post-treatment predicts the need for hospitalization (Gorelick et al. 2004, Kelly et al. 2004). The need for hospital admission among patients with moderate acute asthma, and the need for ICU care in severe acute asthma, was better predicted by asthma severity after 60 minutes of treatment than by the initial assessment in a study of 720 patients seen in 36 Australian EDs (Kelly et al. 2004).

INDICATIONS FOR HOSPITAL ADMISSION

After treatment for 1 to 3 hours in the ED, patients with a poor response to treatment per criteria in Table 3 should be admitted to the hospital. Patients with a marginal response (Table 3) should be assessed individually. The decision to admit or discharge a patient should be made within 4 hours after presentation to the ED (Table 3) (NHLBI 2007).

Table 3 Criteria for Admitting or Discharging the Patient with Acute Exacerbation of Asthma.

Criteria for Discharge	"Grey Zone"*	Criteria for Admission
Good response to treatment (FEV_1 at least 70% of predicted, or PEF at least 70% of personal best)	Marginal response to treatment (FEV_1 40–70% of predicted, or PEF 40–70% of personal best)	Poor response to treatment (FEV_1 less than 40% of predicted, or PEF less than 40% of personal best)
$PaCO_2$ at most 40 mmHg	$PaCO_2$ up to 41 mmHg	$PaCO_2$ at least 42 mmHg
No symptoms	Mild symptoms	Drowsiness, confusion or continuation of other symptoms
Normal exam	Minimal findings	Continued findings such as wheezes
Improvements in lung function, symptoms and signs are sustained for at least 60 minutes	Therapeutic response reaches plateau without further improvement or deterioration	Any deterioration is noted even after initial improvement

*Discharge or admit based on individualized assessment of risk factors for death, ability to adhere to a prescribed regimen, and the presence of asthma triggers in the home, work or school environments.

Indications for Critical Care Hospital Admission

Among patients admitted to the hospital, a subset should be considered for admission to an intensive care unit. Patients at risk for ventilator failure should be considered for intensive care unit admission in addition to those admitted requiring mechanical ventilation initiated in the ED (Table 3) (NHLBI 2007).

DISCHARGE FROM THE ED

Patients may be discharged if after treatment per criteria in Table 3 (NHLBI 2007). Patients should be advised to call their primary care provider or asthma specialist within 3–5 days after discharge from ED for appointment within 1–4 weeks. Data are lacking to show that this action improves outcomes, but this is when the risk of relapse is greatest (NHLBI 2007). Prednisone 40–80 mg daily should be prescribed for 3–10 days as discussed above. ICS should be added (or continued at a higher level for those patients already on an ICS no later than the time of discharge, if not before, to reduce the risk of relapse of acute exacerbation of asthma (Rowe et al. 1999, Sin and Man 2002). If adding ICS anew, a moderate-to-high dose regimen (equivalent of 2 puffs BID of a fluticasone 100 to 250 mcg DPI or 110 to 220 mcg of fluticasone MDI) should be used. SABA should be continued as albuterol/salbutamol 2 puffs four times per day as needed (or equivalent) until the patient connects with their primary care provider or asthma specialist.

Patient Education

The need for treatment in the ED may result from inadequate maintenance therapy and/or insufficient knowledge of how to deal with a worsening of asthma control (NHLBI 2007). Thus, care in the ED should include the practices shown in Table 4. Some of these can be initiated by the primary care or asthma provider if they are available to see the patient in the ED, or can be reinforced and embellished at the first primary care visit post-discharge.

Table 4 Recommendations for Care of the Patient with Acute Exacerbation of Asthma Upon Discharge from the Emergency Department.

Action	Provider
Follow-up	
Advise patient to call their primary care provider or asthma specialist within 3–5 days after discharge from emergency department for appointment within 1–4 weeks	**Emergency Department Provider**
Education	
Teach patient to watch for symptoms and signs of decreasing asthma control with instructions to return	**Emergency Department Provider**
Review details (such as purpose, doses and timing) of asthma medications	**Emergency Department Provider** and/or **Primary Care Provider** (if seen in emergency department or hospital)
Review inhaler technique with patient	**Emergency Department Provider** and/or **Primary Care Provider** (if seen in emergency department or hospital)
Provide patient with an asthma action plan including steps that can reduce exposure to triggers and instructions for monitoring their symptoms and implementing their plan	**Emergency Department Provider** and/or **Primary Care Provider** (if seen in emergency department or hospital)
Medications	
SABA: Continue as albuterol/salbutamol 2 puffs four times per day as needed (or equivalent) until connection with Primary Care Provider	**Emergency Department Provider**
Systemic Corticosteroid: prednisone 40–80 mg daily for 3–10 days	**Emergency Department Provider**
Inhaled Corticosteroid: 2 puffs BID of a moderate-to-high dose agent (equivalent of fluticasone 100 to 250 mcg DPI or 110 to 220 mcg HFA)	**Emergency Department Provider** and/or **Primary Care Provider** (if seen in emergency department or hospital)

DPI=dry-powder inhaler; HFA=hydrofluoroalkane formulation.

PRACTICE AND PROCEDURES

- Upon arrival to the ED or other emergent care area, the patient should be evaluated quickly with focused history, physical examination and diagnostic tests to establish a baseline and for triage to a severity group (Fig. 1).

- Initial treatment in the ED should be simultaneous with the initial evaluation and should include low-flow supplemental oxygen, intravenous fluids and treatment with a SABA (or for those with severe airways obstruction, a SABA-SAMA combination) generally delivered via nebulizer (Fig. 1 and Table 1).
- In the unusual event that signs of impending or actual ventilatory failure are present, plans should be made for intubation under controlled conditions; if this is not possible quickly, BiPAP should be considered as interim therapy in selected patients who are likely to cooperate with mask therapy, or for patients who decline intubation (Fig. 2).
- Further care should include re-evaluation after initial therapy, a second and a third SABA (or combination SABA-SAMA) treatment at 20 minute intervals if indicated, a minimum of prednisone 40 mg and consideration of other treatments if indicated (Table 2 and Fig. 3).
- The case findings should be reviewed relative to criteria for admission to determine the optimal patient disposition (Table 3).
- Planning for discharge of the patient treated emergently for acute exacerbation of asthma should include multiple elements (Table 4).
- Patients and providers should work together to develop a plan to control the patient's asthma, including the avoidance of triggers and adherence to a regimen of ICS and other maintenance or controller medications.
- Patients should be instructed to seek care at their earliest awareness of an increase in symptoms or increased use of rescue medicines.

KEY FACTS

- The initial approach to the patient should be rapid triage to a severity group based on a focused review of the key clinical parameters, while also administering initial therapy. Identification of the severe group is of the utmost importance, because these patients are more likely to benefit from therapy with a combination of inhaled SABA-SAMA and also because a subgroup of these patients might require mechanical ventilator support.

- The patient will require continuous re-evaluation; a key decision point comes after approximately 1 hour of treatment with SABA or SABA-SAMA.
- Low-flow oxygen, intravenous fluids and systemic corticosteroids are also cornerstones of initial therapy.
- For the most severe exacerbations, mechanical ventilatory support may be needed and must be provided as soon as possible, and under controlled conditions with experienced personnel whenever possible. Patients requiring mechanical ventilatory support or who are at high risk for requiring such should be admitted to an intensive care unit.
- Otherwise, the decision to admit or discharge from the ED is based on established admitting criteria.
- Once the exacerbation is under control, care should be directed toward improving asthma control and reducing the risk for exacerbation going forward using ICS and other forms of controller or maintenance therapy.

SUMMARY POINTS

- Acute exacerbations of asthma range from "mild to moderate" to "severe", based on established criteria. Patients in the "severe" group may benefit from SABA-SAMA combination therapy and are at higher risk for requiring support with mechanical ventilation.
- The initial approach to the patient should be rapid triage to a severity group based on a focused review of the key clinical parameters, while also administering initial therapy.
- The patient will require continuous re-evaluation, but especially after approximately 1 hour of treatment.
- Low-flow oxygen, intravenous fluids and systemic corticosteroids are also cornerstones of initial therapy.
- For the most severe exacerbations, mechanical ventilatory support must be provided as soon as possible, and under controlled conditions with experienced personnel whenever possible. Patients requiring mechanical ventilator support or

who are at high risk for requiring such should be admitted to an intensive care unit.

- Otherwise, the decision to admit or discharge from the ED is based on established admitting criteria.
- Further care should be based on improving asthma control and reducing the risk for exacerbation going forward using ICS and possibly other forms of controller or maintenance therapy.

DEFINITIONS OF WORDS AND TERMS

Asthma: A disease characterized by intermittent hyperreactivity of the airways resulting in periods of obstruction to airflow characterized by wheezing, chest tightness, cough and/or dyspnea. The intermittent periods are fully reversible resulting in alternative periods of normal airflow and the absence of symptoms.

Albuterol/salbutamol: A commonly-used inhaled SABA, called albuterol in the USA and salbutamol outside the USA.

Bi-level positive airway pressure (BiPAP): A mechanical system for providing mechanical assistance to ventilation that does not require placement of a plastic endotracheal tube. This form of treatment may temporize patients with acute exacerbation of asthma until their airways obstruction responds to therapy or endotracheal intubation is available under controlled conditions with experienced personnel.

Corticosteroids: A group of medications that resemble the hormone produced normally by the adrenal cortex and having anti-inflammatory effects and effects on glucose homeostasis and sodium homeostasis.

Exacerbation: An episodic worsening of a chronic disease. In asthma, this may be subacute and lasting for days to weeks with a presentation usually to a practitioner's office, or acute lasting for minutes to hours prior to presentation to an ED or similar emergent-care environment.

Forced expiratory volume at 1 second (FEV_1): The volume of air, measured using a spirometer, moved during the first second of forced exhalation; the procedure of measuring a forced exhalation with a spirometer is called "spirometry" and produces other results such as the forced vital capacity (FVC).

Global Initiative for Asthma (GINA): An initiative sponsored by a consortium of health-care companies to promote better asthma outcomes, available at www.ginasthma.org

Ipratropium: A commonly-used inhaled SAMA.

Metered-dose inhaler (MDI): The common device designed to deliver a measured dose of inhalational medicine in a minimal volume of excipient. These devices require hand-breath coordination, the need for which can be minimized by using a valved-holding chamber or "spacer".

National Asthma Education and Prevention Program Expert Panel Report 3 (NAEPP-3): An initiative sponsored by the National Heart, Lung and Blood Institute (NHLBI) of the National Institutes of Health, Department of Health and Human Services, USA) to promote better asthma outcomes, available at www.nhlbi.nih.gov/guidelines/asthma/asthgdln.pdf

Peak expiratory flow rate (PEF): The maximum flow, measured using a peak-flow-meter, exhaled during a forced exhalation.

Salbutamol: A commonly-used inhaled SABA, called albuterol in the USA.

Short-acting β_2-adrenergic agonist (SABA): An inhaled drug with bronchodilator properties based on stimulation of the beta-adrenergic nervous system with a relatively short (less than 8 hours) duration of action.

Short-acting muscarinic antagonist (SAMA): An inhaled drug with bronchodilator properties based on inhibition of the cholinergic nervous system with a relatively short (less than 8 hours) duration of action.

LIST OF ABBREVIATIONS

BiPAP	:	bi-level positive airway pressure
ECG	:	electrocardiogram
ED	:	emergency department
FEV_1	:	forced expiratory volume at 1 second
GINA	:	Global Initiative for Asthma
ICS	:	inhaled corticosteroids
MDI	:	metered-dose inhaler

NAEPP-3	:	National Asthma Education and Prevention Program Expert Panel Report 3
PEF	:	peak expiratory flow rate
SABA	:	short-acting β_2-adrenergic agonist
SAMA	:	short-acting muscarinic antagonist

REFERENCES

Bateman, E.D., S.S. Hurd, P.J. Barnes, J. Bousquet, J.M. Drazen, M. FitzGerald, P. Gibson, K. Ohta, P. O'Byrne, S.E. Pedersen, E. Pizzichini, S.D.Sullivan, S.E. Wenzel and H.J. Zar. 2008. Global strategy for asthma management and prevention: GINA executive summary. Eur Respir J 31: 143–78.

Bowler, S.D., C.A. Mitchell and J.G. Armstrong. 1992. Corticosteroids in acute severe asthma: effectiveness of low doses. Thorax 47: 584–7.

Camargo, C.A. Jr., D.M. Gurner, H.A. Smithline, R. Chapela, L.M. Fabbri, S.A. Green, M-P Malice, C. Legrand, S. Balachandra Dass, B.A. Knorr and T.F. Reiss. 2010. A randomized placebo controlled study of intravenous montelukast for the treatment of acute asthma. J Allergy Clin Immunol 125: 374–80.

Camargo, C.A.Jr., C.H. Spooner and B.H. Rowe. 2003. Continuous vs. intermittent beta agonists in the treatment of acute asthma. Cochrane Database Syst Rev 4: CD001115.

Dhuper, S., A. Chandra, A. Ahmed, S. Bista, A. Moghekar, R. Verma, C. Chong, C. Shim, H. Cohen and S. Choks. 2011. Efficacy and cost comparisons of bronchodilator administration between metered dose inhalers with disposable spacers and nebulizers for acute asthma treatment. J Emerg Med 40: 247–55. doi: 10.1016/j.jemermed.2008.06.029

Dockhorn, R.J., R.A. Baumgartner, J.A. Leff, M. Noonan, K. Vandormael, W. Stricker, D.E. Weinland and T.F. Reiss. 2000. Comparison of the effects of intravenous and oral montelukast on airway function: a double blind, placebo controlled, three period, crossover study in asthmatic patients. Thorax 55: 260–5.

Edmonds, M.L., C.A. Jr. Camargo, C.V. Jr. Pollack and B.H. Rowe. 2003. Early use of inhaled corticosteroids in the emergency department treatment of acute asthma. Cochrane Database Syst Rev 3: CD002308.

Emerman, C.L. and R.K. Cydulka. 1995. A randomized comparison of 100-mg vs 500-mg dose of methylprednisolone in the treatment of acute asthma. Chest 107: 1559–63.

Engel, T., A. Dirksen, L. Frølund, J.H. Heinig, U. Gerner Svendsen, B. Klarlund Pedersen and B. Weeke. 1990. Methylprednisolone pulse therapy in acute severe asthma: a randomized, double-blind study. Allergy 45: 224–30.

Gorelick, M.H., M.W. Stevens, T.R. Schultz and P.V. Scribano. 2004. Performance of a novel clinical score, the Pediatric Asthma Severity Score (PASS), in the evaluation of acute asthma. Acad Emerg Med 11: 10–8.

Jónsson, S., G. Kjartansson, D. Gíslason and H. Helgason. 1988. Comparison of the oral and intravenous routes for treating asthma with methylprednisolone and theophylline. Chest 94: 723–6.

Kelly, A.M., D. Kerr and C. Powell. 2004. Is severity assessment after one hour of treatment better for predicting the need for admission in acute asthma? Respir Med 98: 777–81.

Krishnan, J.A., S.Q. Davis, E.T. Naureckas, P. Gibson and B.H. Rowe. 2009. An umbrella review: corticosteroid therapy for adults with acute asthma. Am J Med 122: 977–91.

Lazarus, S.C. 2010. Emergency treatment of asthma. N Engl J Med 363: 755–648.

Liu, M.C., L.M. Dubé and J. Lancaster. 1996. Acute and chronic effects of a 5-lipoxygenase inhibitor in asthma: a 6-month randomized multicenter trial. J Allergy Clin Immunol 98: 859–71.

Marquette, C.H., B. Stach, E. Cardot, J.F. Bervar, F. Saulnier, J.J. Lafitte, P. Goldstein, B. Wallaert and A.B. Tonnel. 1995. High-dose and low-dose systemic corticosteroids are equally efficient in acute severe asthma. Eur Respir J 8: 1435.

Mohammed, S. and S. Goodacre. 2007. Intravenous and nebulised magnesium sulphate for acute asthma: systematic review and meta-analysis. Emerg Med J 24: 823–30.

Moorman, J.E., R.A. Rudd, C.A. Johnson, M. King, P. Minor, C. Bailey, M.R. Scalia and L.J. Akinbami. 2007. National surveillance for asthma—United States, 1980–2004. MMWR Surveill Summ 56: 1–54.

National Heart, Lung, and Blood Institute, National Asthma Education and Prevention Program. Expert Panel Report 3: guidelines for the diagnosis and management of asthma: full report 2007. (Accessed January 18, 2011 at http://www.nhlbi.nih.gov/guidelines/asthma/asthgdln.pdf)

Parameswaran, K., J. Belda and B.H. Rowe. 2000. Addition of intravenous aminophylline to beta2-agonists in adults with acute asthma. Cochrane Database Syst Rev 4: CD002742.

Plotnick, L.H. and F.M. Ducharme. 2000. Combined inhaled anticholinergics and beta2-agonists for initial treatment of acute asthma in children. Cochrane Database Syst Rev 4: CD000060.

Qureshi, F., A. Zaritsky, C. Welch, T. Meadows and B.L. Burke. 2005. Clinical efficacy of racemic albuterol vs. levalbuterol for the treatment of acute pediatric asthma. Ann Emerg Med 46: 29–36.

Rodrigo, C. and G. Rodrigo. 1998. Therapeutic response patterns to high and cumulative doses of salbutamol in acute severe asthma. Chest 113: 593–8.

Rodrigo, G., C. Pollack, C. Rodrigo and B.H. Rowe. 2006. Heliox for nonintubated acute asthma patients. Cochrane Database Syst Rev 4: CD002884.

Rodrigo, G.J. and J.A. Castro-Rodriguez. 2005. Anticholinergics in the treatment of children and adults with acute asthma: a systematic review with meta-analysis. Thorax 60: 740–6. (Errata: Thorax 2008; 63: 1029, 2006; 61: 274, 458).

Rodrigo, G.J. and C. Rodrigo. 2002. Continuous vs intermittent beta-agonists in the treatment of acute adult asthma: a systematic review with meta-analysis. Chest 122: 160–5.

Rowe, B.H., G.W. Bota, L. Fabris, S.A. Therrien, R.A. Milner and J. Jacono. 1999. Inhaled budesonide in addition to oral corticosteroids to prevent asthma relapse following discharge from the emergency department: a randomized controlled trial. JAMA 281: 2119–26.

Rowe, B.H. and C.A. Camargo Jr. 2008. The role of magnesium sulfate in the acute and chronic management of asthma. Curr Opin Pulm Med 14: 70–6.

Rowe, B.H., D.C. Voaklander, D. Wang, A. Senthilselvan, T.P. Klassen, T.J. Marrie and R.J. Rosychuk. 2009. Asthma presentations by adults to emergency departments in Alberta, Canada: a large population-based study. Chest 135: 57–65.

Sin, D.D. and S.F. Man. 2002. Low-dose inhaled corticosteroid therapy and risk of emergency department visits for asthma. Arch Intern Med 162: 1591–5.

Soroksky, A., D. Stav and I. Shpirer. 2003. A pilot prospective, randomized, placebo-controlled trial of bilevel positive airway pressure in acute asthmatic attack. Chest 123: 1018–25.

Thill, P.J., J.K. McGuire, H.P. Baden, T.P. Green and P.A. Checchia. 2004. Noninvasive positive pressure ventilation in children with lower airway obstruction. Pediatr Crit Care Med 5: 590.

9

Nurse-led Home Visits for Difficult Asthma

Pippa Hall[1,a,*] **and** **_Andrew Bush_**[1,b]

ABSTRACT

In up to 50% of patients with 'problematic, severe asthma' a home-visit allows the identification of remediable factors. This visit allows the nurse to assess the home environment and identify areas for improvement. It also identifies which patients require more invasive investigations. The key areas of assessment are on-going allergen exposure, in particular House Dust Mite (HDM) and pets, and whether any steps are in place to effectively minimise exposure. Exposure to cigarette smoke from parents or other family members within the home as evidenced by the characteristic odour and the presence of ashtrays or actual smoking is observed. Adherence to medications prescribed is determined by checking prescription pick up records, and especially that appropriate in-date medications are available and easily accessible within the home. Finally, the home visit allows an opportunity to discuss any psychosocial issues; often it only becomes apparent on the home-visit that there are contributory psychosocial issues which can be discussed and recommendations made such as referral to psychology for management. Once identified, such contributory factors can be addressed hopefully leading to improved asthma control and obviating the need for more toxic and 'beyond guideline' treatment which otherwise would have been prescribed.

[1]Royal Brompton Hospital, Sydney Street, London, SW3 3NP, United Kingdom.
[a]Email: P.Hall@rbht.nhs.uk
[b]Email: a.bush@rbht.nhs.uk
*Corresponding author

List of abbreviations after the text.

INTRODUCTION

When children are sent for the first time to a tertiary referral centre, they are considered to come under the umbrella term of 'problematic, severe asthma' (Bush et al. 2008, Hedlin et al. 2010). The purpose of evaluation is to determine which children have truly got 'severe, therapy resistant asthma' requiring experimental, high risk treatments (and these are the minority) and who can be managed more appropriately in other ways. This chapter will in particular focus on the role of the respiratory nurse in this process, set in the context of the overall evaluation.

PATTERNS OF DIFFICULTY WHICH TRIGGERS REFERRAL TO SPECIALIST CARE

The different problems which trigger referral can be broken down into one or more of:

1. Persistent (most days, for at least 3 months) chronic symptoms (short-acting β-2 agonist use > three times/week) of airways obstruction despite high dose inhaled corticosteroids (ICS) (Beclomethasone equivalent 800 mcg/day) and trials of add-on medications. The ICS threshold is arbitrary. It should be noted that there is a poor correlation between symptoms and parental administration of β-2 agonists.
2. Type 1 brittle asthma [Ayres et al. 1998] (chaotic swings in peak flow). Definitions and virtually all data are in adults.
3. Recurrent severe asthma exacerbations that have required:
   - *either* at least one admission to an intensive care unit
   - or at least two hospital admissions requiring intravenous medication/s
   - *or* > 2 courses of oral steroids during the last year, despite the above therapy.

(The choice of numbers throughout is arbitrary).

4. Type 2 brittle asthma [Ayres et al. 1998] (sudden and catastrophic attack on the basis of apparently good control); again, most data come from adults.

5. Persistent airflow limitation (PAL): post oral steroid, post-bronchodilator Z score < −1.96 for first second forced expired volume (FEV_1) using appropriate reference populations.
6. The necessity of prescription of alternate day or daily oral steroids.

THE ENTRY LABEL: 'PROBLEMATIC SEVERE ASTHMA'

In our practice, children with *problematic, severe asthma* are severely disabled. We studied 71 children (35 male), 21 of whom were using regular oral steroids [Bossley et al. 2009]. The mean dose of fluticasone equivalent was 1 mg/day, range 0.5 to 3 mg/day. They had a median of 2 admissions to hospital, range 0–21, and 12 were ventilated on at least one occasion. Mean first second forced expired volume in one second (FEV_1) was 76% (range 33–125), and despite prescribed medication, median bronchodilator reversibility was 14% (range 12–106). 34% had persistent airflow limitation, defined here as FEV_1 < 75% predicted despite prednisolone and high dose β-2 agonists. 97% had an asthma control test < 20. Median FeNO was 52 ppb (range 5–171, normal < 25). Atopy was common, with more than 50% being skin prick test (SPT) positive to house dust mite (HDM), grasses, cat and dog. Food sensitivity at least as judged by SPT was common (peanut 25%, egg and milk 5–10%).

Problematic severe asthma comprises four groups: 'not asthma at all' (wrong diagnosis); 'asthma plus' a significant comorbidity; and the two categories in which the role of the nurse is pivotal, 'difficult asthma' in which basic management has not been got right; and 'severe, therapy resistant asthma' in which despite every effort, conventional management has failed (Fig. 1).

Not asthma at all The differential is beyond the scope of this chapter. If the another diagnosis made, then no further asthma evaluation is performed.

Asthma plus Common co-morbidities have been reviewed in detail elsewhere (De Groot et al. in 2010). They include gastro-oesophageal reflux, rhinosinusitis, dysfunctional breathing, obesity and food allergy.

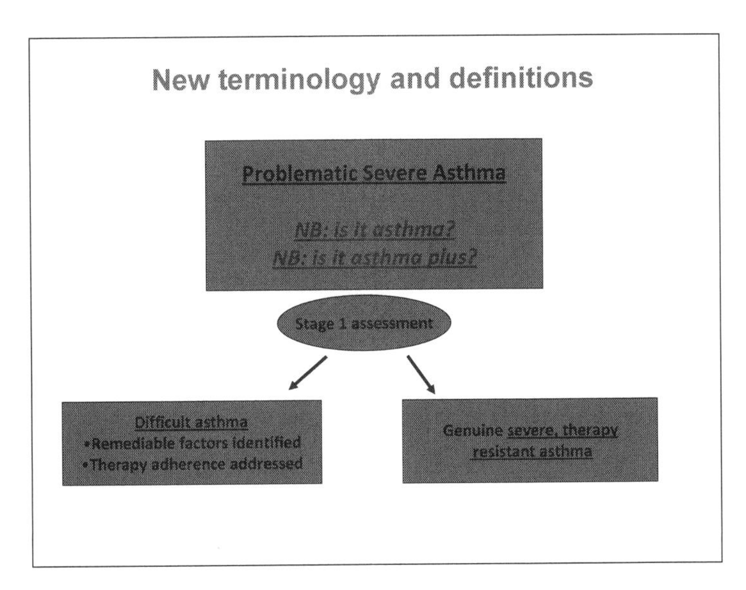

Figure 1 New terminology and definitions.

'Difficult asthma' and 'Severe, therapy resistant asthma' There is ample evidence from the literature that getting the basics right leads to much apparently severe asthma being revealed as no such thing. Three recent studies illustrate this well. The first was a trial of add-on therapy in children who had persistent asthma despite moderately high dose ICS and long acting β-2 agonists [Struck et al. 2008]. 292 children were assessed for entry, but only 55 were randomised, and the study, which was negative, was futile for want of power. The key reasons for exclusion were non-adherence to treatment or that the child could not be shown to have asthma. The other two studies were related to the use of exhaled nitric oxide (FeNO) to improve asthma control [Szefler et al. 2008, de Jongste et al. 2009]. In a study of inner city asthma, by the time the basics had been got right asthma control was so much better that there was really little scope for further improvements by measuring FeNO [de Jongste et al. 2009]. In the second trial, FeNO telemonitoring to guide asthma therapy was compared to a standard regime. There was intensive input and monitoring in both limbs of the study, and both groups improved equally.

The remainder of this chapter describes how we separate out children with 'Difficult asthma'. The further investigation and management of severe, therapy resistant asthma is described in detail elsewhere [Bush and Saglani 2010].

THE IMPORTANCE OF NURSING INPUT

The role of the nurse in asthma management is to help children and families keep control of their asthma while on the minimum amount of treatment. Improved baseline control and the prevention of asthma attacks involve identifying and avoiding triggers, patient education and effective management plans, and ensuring adequate adherence to medication. A home visit carried out by a specialist nurse can be invaluable. It can also help to identify which patients have severe therapy resistant asthma in which every conventional management option has failed and those patients with asthma that will improve if remediable issues are identified and eliminated (discussed in more detail below). Furthermore following a home visit an effective management plan can then be implemented to help avoid the need for further investigations and escalations of treatment (Bracken et al. 2009).

Assessment of patients with severe asthma is usually carried out by doctors in the hospital environment. Useful information can be gained this way, however it is often incomplete due to time constraints (Ogden et al. 2004) and reluctance of the family to disclose sensitive information to a doctor (Barry et al. 2001). There is growing evidence supporting the benefit of nurse-led home visits. Nurse-led home visits have been shown to be of benefit in helping patients with asthma develop self-management plans (Horner 2006), to minimise exposure to home environmental asthma triggers and to provide psychological support.

The aim of the home visit is to obtain further information about the patient and family that has not been discovered or disclosed in the hospital setting. As a result, more tailored advice and medical management can be delivered to the patient and family. Often drug treatment can be altered and more toxic 'beyond guideline' treatment can be avoided. Although it is appreciated that much of the information collected can theoretically be gained from direct questioning within the clinical setting, in fact research has shown that further more accurate and useful information can be gained from a home visit (Bracken et al. 2009). This chapter will outline the main areas such as allergen exposure, tobacco smoke, adherence and psychosocial issues that the nurse-led home visit should focus on, ultimately aiming to solve the puzzle of why asthma is severe (Fig. 2).

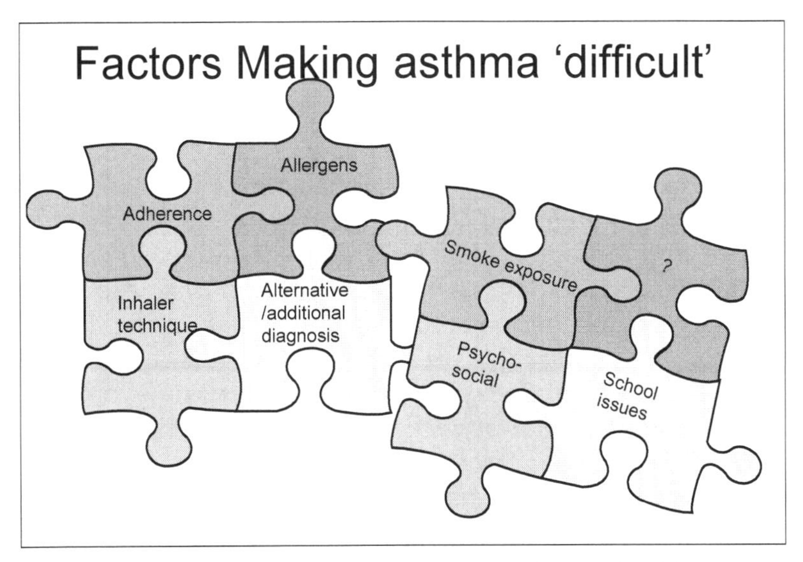

Figure 2 Solving the puzzle; factors that can make asthma 'difficult'.

PRACTICE AND PROCEDURES

The home visit should be arranged directly between the family and the nurse at a mutually convenient time. The reason for the visit should be discussed with the family, and it should be stressed that it is to help identify anything that may be in the home or happening at home which could be improved upon which in turn could improve asthma control. The visit is an example of working in partnership, not the nurse trying to catch the family out in some way. It may be valuable to consider doing a joint visit with another professional involved with the child's asthma such as the local community nurse if appropriate.

There are limitations and challenges of visiting the home. It is imperative that home visits are only carried out with appropriate safety measures in place. Fixing a time may be difficult if both parents work outside the home. Scheduling is easier if the child does not need to be present. The nurse must appreciate that families have many competing demands for their time and a visit from the asthma nurse may not be of high priority for every family. However, if such a visit is of low priority in a family with a child with really severe asthma, this calls into question their commitment to getting the asthma treatment right. The home can be very different to the clinical setting and the nurse needs to adapt to manage the home environment. The nurse must not forget that she is a visitor in someone else's home and show respect for this.

Giving advice that may not be pleasing to a family can be difficult in the clinical environment, however when the nurse is in someone's home it can be even harder. Advising a parent that their smoking habits are contributing to their child's poor asthma control can be difficult listening for a parent who is unwilling or unable to cut back or quit. Advising a family that they should be re-housing their pets can be equally difficult.

The whole process of carrying out a home visit is time consuming. Visits are likely to take between 2 and 4 hours depending on individual needs of the family. Travel time can also be extensive. Furthermore time is spent liaising with GP practices when checking adherence, schools to ensure suitable management plans and finally discussing the findings within the multidisciplinary team. Following up patients and ensuring interventions recommended are being adhered to also add to the over-all time invested.

GENERAL HOME ENVIRONMENT

The nature of the home and its immediate surroundings may be impacting on the asthmatic child. Visiting can elicit information that can be of value. There are several areas that should be considered as potential factors that could be contributing to poor asthma control.

Visiting the home gives insight into the type of property the family lives in. If the property is a flat such information as to what floor is it on can be worthy of note? Do other residents in the block smoke in the communal areas such as lifts and stairwells? Are there lots of pets living in the same block and frequenting the communal areas? It is useful to know if the home is owner occupied, privately rented or council rented. Information about type of ownership is paramount in determining how easy necessary modifications can be made within the home if needed.

The immediate surroundings of the home may be of relevance to the asthma patient. First hand knowledge that a home is situated on a very busy road will be of help in offering practical advice in regards to the usual recommendations of the benefits of opening windows and promoting good ventilation within the home, as it may be more appropriate to keep windows closed. Another area to consider is the social demographics of neighbourhood. It has been widely reported that neighbourhood factors such as greater violence and lower socioeconomic status has an adverse affect on asthma morbidity and increased allergic responses (Watson et al. 1996).

Text Box: Useful checklist for home environment

The Home: Flat ☐ House ☐

If a flat: What Floor? ☐ Lift available? ☐ Stairs? ☐

Description of communal area:

Description of immediate surroundings:

Number of bedrooms ☐ Own bedroom ☐

Family/household numbers ☐

Overall impression of home organisation:

Once inside the home it is possible to make a subjective judgement as to how organised it is. A lack of basic organisation is cause for concern. If parents of a child with 'severe' asthma have forgotten about the nurse visit it could be an indication that other more important things may get missed such as giving medication. If parents continue to watch the television during the visit or continue to receive calls on their mobile it could indicate lack of parental concern or perception of the importance of the home-visit and severity of their child's asthma. Reasons for this can be further explored if appropriate.

Visiting the home allows an assessment of who else lives there. Many households will have grandparents, cousins, step siblings, cousins and friends all living in the same place. Often residents are only temporary and it's not until a home visit is carried out that information is gathered. The number of bedrooms in the home and whether or not the child has to share a bedroom can be observed. During the home visit over crowding within the home can be identified and appropriate action can be made if required such as formal recommendations for re-housing.

ALLERGENS

Exposure to indoor aeroallergens such as House Dust Mites (HDM), pets and moulds at a level insufficient to cause acute symptoms will lead to increased airway inflammation, bronchial responsiveness and steroid resistance (Sulakvelidze et al. 1998). High allergen exposure in the home and allergic sensitization is associated with acute exacerbation in children (Murray et al. 2006). There is also some evidence showing that allergens can worsen asthma by non-IgE –mediated mechanisms (Green et al. 2002), therefore minimising exposure in the home is paramount in improving asthma control and minimising treatment. Prior to visiting the home it is helpful for children to have been allergy tested for the common aeroallergens. This should be done both by skin prick testing and taking a blood sample and testing for specific IgE (Frith et al. in press). The nurse should have access to the results so that advice can be specific during the visit (Fig. 3).

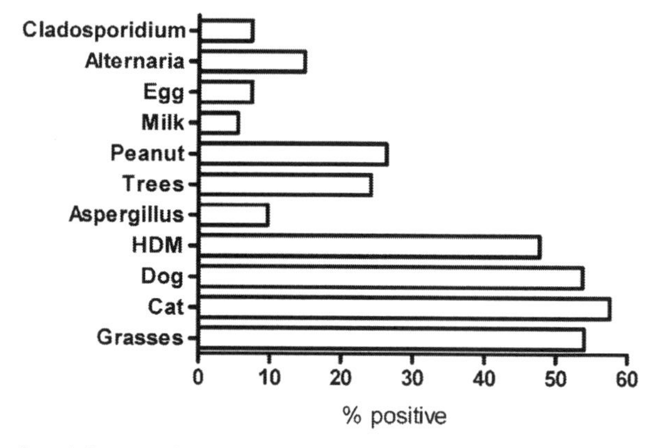

Figure 3 Results of skin prick tests to common aeroallergens and foods in our group of patients.
A positive response was defined as a wheal ≥3mm in children ≥6 years of age and ≥2mm in children < 6 years. Results are expressed as the percent positive of those tested.
Reproduced from: Bracken M., L. Fleming, P. Hall, N. Van Stiphout, C. Bossley, E. Biggart, N. Wilson and A. Bush. 2009. The importance of home visits in children with problematic asthma. Arch Dis Child 94: 780–4 (with permission).

Text Box: Common aeroallergens and checklist to record sensitivity results (useful information prior to home visit). Other allergens may be tested depending on circumstances, for example guinea pig if this is a family pet

Cats	Dogs
Aspergillus	Grass
Trees	Cladosporium
House Dust Mites (HDM)	Alternata Alternaria
Cockroach	Penicillium
Milk	Egg
peanut	

Children who are sensitised to HDM and pets will have previously been given allergen avoidance advice in the clinical setting. However at the home visit in most cases steps taken to minimise allergen exposure were sub-optimal (Bracken et al. 2009). The role of allergen avoidance in the management of asthma is controversial and it has been reported that HDM avoidance is of no value whatever (Gotzsche and Johnson 2008). However, many studies cited did not actually achieve a reduction in HDM levels, were very short term, or were carried out in patients with mild disease, yet this is taken as evidence by some that reduction of allergen load is not beneficial. There are also no data in children with severe asthma. Many children are sensitised to multiple allergens and research has shown that when reducing allergen load in these patients there are benefits (Horner 2006). For this reason it is believed that benefit can be achieved by reducing allergens in the home and effort should be made by the nurse to facilitate this being achieved.

Effective ways to reduce HDM allergens include allergen-proof encasings fitted to the mattress and pillow, thorough vacuum cleaning with a high efficiency particulate air (HEPA) filter and washing all bedding at 60 degrees once a week.

When giving advice on methods to minimise allergen exposure in the home, the nurse should be aware of the financial and practical constraints that are individual to a particular household. Families on low income and living in social housing will be less able to purchase appropriate equipment such as a HEPA filter vacuum cleaner and expensive protective bedding. The removal of carpets and their replacement with a more suitable floor covering can put a financial burden on some families. Where appropriate the nurse can help the family apply for grants to help cover the costs. The suggestion that a much loved family pet should be removed from the home may not be met with much enthusiasm.

Useful Checklist: Allergen exposure

Patient allergy results from hospital SPT's and sIgE:				
Pets in the home □	Details:			
Evidence of moulds □	If yes:	Walls □		
		Windows □		
		Damp smell □		
HDM Avoidance measures	Mattress Cover □			
	Pillow protection □			
	Linen regularly washed at 60□C □			
	Flooring:	Carpet □	Laminate/wood	□
	Windows:	Blinds □	Curtains	□
	Evidence of damp dusting □			
	Presence of soft toys □			
	HEPA filter vacuum cleaner □			
	Leather sofas □	Fabric sofas □		
Presence of air fresheners &/or other potential irritants □		details:		

Research has shown that many families do not undertake HDM reduction measures within the home despite positive allergy testing and advice from health professionals to do so. Of 71 severe asthmatic children visited at home 31 had positive skin prick tests to HDM. In 35% of them there were no HDM avoidance measures in place, despite prior advice. 48% had some avoidance measures in place and 16% had reasonable measures in place. For patients where HDM avoidance measures are sub-optimal advice can be re-enforced at the time of the home visit. In the same group 30 had a cat, dog or both, 17 of whom had a positive SPT to their pet. Although every effort should be made before the home visit to encourage the removal of the pet from the home, if this has not been achieved advice can be given to help minimise pet exposure such as not allowing the pet upstairs or keeping it outside. Also, hands should be washed after any contact with the pet. If the pet is successfully removed from the home families must be advised that allergen levels in settled dust decrease to those seen in homes without cats only over 4 to 6 months, and an immediate benefit will not be seen (Wood et al. 1989). Levels will decrease faster if further environmental control measures are undertaken such as the removal of carpets and soft furnishings. It should be recommended further that as cat allergens could remain in mattresses for many years after the cat has been removed, the mattress should be replaced.

Key Facts of allergen exposure

- Most 'problematic severe asthmatic' patients visited are atopic.
- About one third of patients are sensitised to HDM however only half of those have reasonable HDM avoidance measures in place and a third did no avoidance measures at all.
- Half of such children with pets in the home are sensitised to their pet.
- Approximately one third of 'problematic severe' asthmatics have on-going allergen exposure in the home.
- Allergen avoidance measures can be costly and the home visit allows advice to be tailored to individual needs.

SMOKING

Active cigarette smoking has a detrimental effect on the lungs and causes steroid resistance (Chalmers et al. 2002). It is likely that passive cigarette smoke exposure can have similar effects. Passive smoke exposure is common in asthmatic children; 25% of children with severe asthma were exposed to cigarette smoke (Bracken et al. 2009). The prevalence of active smoking amongst children with asthma is unknown but is likely to be high. Measuring salivary cotinine levels identifies those children being exposed or are active smokers. Although home visiting cannot be used to identify which children have parents or other family members who smoke, it can identify those parents who only smoke outside. Household indoor smoking is often apparent immediately from the characteristic smell. However, as visiting is always pre-arranged and families have plenty of prior warning suspicion should be raised if there is an overwhelming smell of air fresheners (which may themselves be detrimental to asthma control). Other more obvious evidence of parental smoking is the presence of ashtrays both indoors and outdoors, used cigarette butts in the garden and the smell of cigarette smoke on the family's clothes and breath. There are of course the few parents who will smoke a cigarette while the nurse is present!

Text box: Checklist about smoking habits

Results of urinary cotinine test:					
Do parents smoke?	☐	if yes: Mother ☐	Father ☐	Other	☐
Does child smoke	☐				
Evidence of smoking in the home	☐	Characteristic odour			☐
		Presence of ashtrays			☐
		Actual smoking observed			☐
Evidence of outside smoking	☐				

Educating families on the harm passive smoke exposure is having on their child can be further re-enforced at the home visit. Time can be spent while at the home exploring with the family ways to reduce smoke exposure. More tailored and practical advice can be given when the nurse can fully appreciate the type of domestic arrangements. For parents who wish to stop smoking, advice and help should be provided.

Key Facts of smoke exposure

- 25% of asthmatic children are likely to be exposed to cigarette smoke.
- Urinary cotinine levels should be measured while the home visit helps determine the extent of smoke exposure from observation

of active smoking, presence of ashtrays and/or a characteristic odour.
- Time can be spent exploring with the family ways to minimise smoke exposure.

ADHERENCE TO TREATMENT

Poor adherence to prescribed medication is common in children and adolescents with chronic illness (Dean et al. 2010). A recent study carried out found that medication issues including adherence, unsuitable device and poor inhaler technique contributed to poor symptom control in 48% of a group of patients being investigated for 'severe' asthma (Bracken et al. 2009).

Before visiting a child at home, assessments of inhaler technique and basic understanding of treatment regimes can be made in the hospital settings. An accurate list of medications and exact doses prescribed by the hospital should also be compiled prior to visiting at home so that it can be compared to what is actually happening at home. Hospital and GP prescriptions can be checked and calculations made to determine the uptake of drug collection. In the same study looking at the importance of home-visits in a cohort of 71 severe asthmatic children only 43% of patients had above 80% prescription pick-up and 30% had below 50% pick-up rate. This information can provide very useful data in the families where pick-up is poor—clearly adherence to the basic treatment will be poor. However even if pick-up rates are high, poor adherence cannot be excluded, for other reasons that can be ascertained on the home visit.

On visiting the home the nurse should ask to see where all the asthma medication is stored. The location of the medications within the home can give useful information. Is there a salbutamol or terbutaline inhaler easily at hand, and is the delivery device appropriate? Can all medications be found easily? There may be a salbutamol pMDI by the patient's bed but no spacer. In the clinical setting it is possible for a child to give a perfect demonstration of correct inhaler technique but it is then apparent that the most appropriate device is not being used at home. If a home has more than one family member with asthma it is possible for the treatment regimes to become mixed. Time should be spent simplifying treatment regimes as much as possible.

Expiry dates on all medication should be checked. 11% of patients had expired medication in the home (Bracken et al. 2009). This would lead the nurse to question adherence and explore with the family why medication is left beyond the 'use by' date. Education in the importance of not using out of date medication can be given and Inappropriate stockpiling of medication can be identified. If prescriptions are being collected regularly but not being used time can be spent exploring why this is happening. One reason may be if parents were not supervising their child taking treatment. A recent study found that by the age of 11, 50% of children were taking ICS unsupervised (Orrell-Valente et al. 2008).

Text box: Adherence to treatment checklist

Current Medication list:			
For each prescribed drug:	Available ⍰		
	In date ⍰		
	Spares available ⍰		
Appropriate inhaler devices ⍰	Re-check inhaler technique ⍰		
Evidence of poor supervision ⍰			
Location of medications:			
Appropriate knowledge of medication regime ⍰			
Percentage of medication uptake from prescription records:		>80%	⍰
		50- 80%	⍰
		> 50%	⍰
Evidence of inappropriate stockpiling ⍰			

Time spent in the home provides the perfect opportunity for the nurse to ensure that the family has an appropriate understanding of the treatment regime. It is important that what each drug does is explained to the family. At home, the nurse is more likely to do this well.

Key Facts on Adherence

- Approximately 50% of children have medication issues that contribute to poor asthma control.
- Treatment issues are commonly related to asthma severity.
- Children who are not supervised while taking their medication are unlikely to be taking their treatment regularly.
- Ensuring that medications are being administered correctly and regularly reduces the need for more toxic 'beyond guideline' treatment.

PSYCHOSOCIAL ISSUES

There is a wide range of psychosocial morbidity in patients with asthma (Richardson et al. 2008, Heaney et al. 2006). We do not try to find out if stress is causing the asthma, or asthma is causing the stress but treat both on merit. The importance of acute and chronic stress as a trigger of asthma exacerbations is well recognised (Sandberg et al. 2000). Prior to the home visit the nurse will have had the opportunity to explore such issues with the family. Helpful tools include the Asthma Control Test (ACT) and Juniper questionnaire (Juniper et al. 1996) which can help gauge the level of impact asthma plays in daily activities such as exercise limitation, sleep quality and time off school and overall quality of life. A clinical psychologist should help interpret the results (Fig. 4). Although such tools can

Psychosocial Factor	Number (%)
Child anxiety / depression / other psychological issue	20 (28)
Parental anxiety / depression / other psychological issue	20 (28)
Perception / dysfunctional breathing	11 (15)
School issues	4 (6)

Figure 4 Psychosocial factors identified in our group of patients. More than one could be assigned to each subject.
Reproduced from: Bracken M., L. Fleming, P. Hall, N. Van Stiphout, C. Bossley, E. Biggart, N. Wilson and A. Bush. 2009. The importance of home visits in children with problematic asthma. Arch Dis Child 94: 780-4 (with permission).

be useful in eliciting information parents and children appear to feel more comfortable at home and are much more likely to disclose personal information during the home visit, especially if the child is not present. The relationship between the nurse and the family is paramount in the success of uncovering any issues that may be contributing to poor asthma control. If psychosocial issues are likely contributors to the severity appropriate referrals can be made to clinical psychologist and local Children and Adolescent Mental Health Services (CAMHS).

Textbox: Checklist for Psychosocial issues

Previous psychosocial issues identified	☐
If yes details:	
Psychosocial issues discussed at time of home visit:	
Appropriate perception of asthma severity	☐
Referral to psychology	☐

MULTIDISCIPLINARY TEAM INVOLVEMENT

Following the home visit the findings are discussed within the multidisciplinary team. Referrals can be made, for example to clinical psychology or CAMHS. For children where remedial factors had been identified and improved upon further investigations or treatment escalation are avoided. When interventions are recommended it is important to check that any are actually being made by the family, for example re-checking GP prescription records after addressing poor adherence.

FINAL OUTCOMES

We could [Bracken et al. 2009] not even retrospectively to find a way of selecting problematic severe asthmatics who did not need a home visit. As a consequence of the assessment 55% of the children had remedial factors identified (Fig. 5) and this meant that further escalation of treatment was avoided.

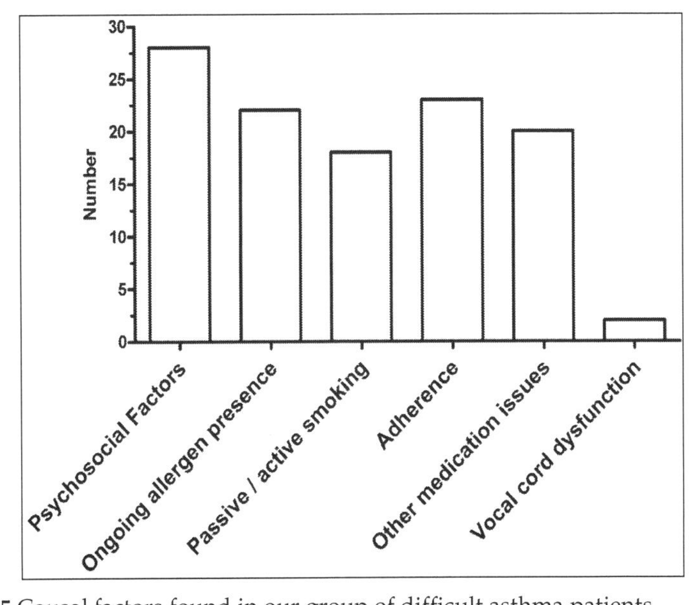

Figure 5 Causal factors found in our group of difficult asthma patients. More than one factor could be found per child
Reproduced from: Bracken M., L. Fleming, P. Hall, N. Van Stiphout, C. Bossley, E. Biggart, N. Wilson and A. Bush. 2009. The importance of home visits in children with problematic asthma. Arch Dis Child 94: 780–4 (with permission).

SUMMARY

- Home visits can determine which patients have factors contributing to asthma severity (Fig. 6), which if identified can obviate the need for more invasive investigations and 'beyond guideline' treatment, while improving quality of life.
- In the clinic setting most parents only report smoking outdoors a home visit may identify households where smoke was detectable indoors.
- Allergen avoidance measures can be expensive and the home visit gives an opportunity to identify the key areas for effective intervention.
- Home visits can identify psychosocial problems that otherwise may not be discovered if consultations are limited only to the hospital setting.
- There are limitations to home visits; however the benefits outweigh the costs in children whose asthma is seemingly difficult to treat.

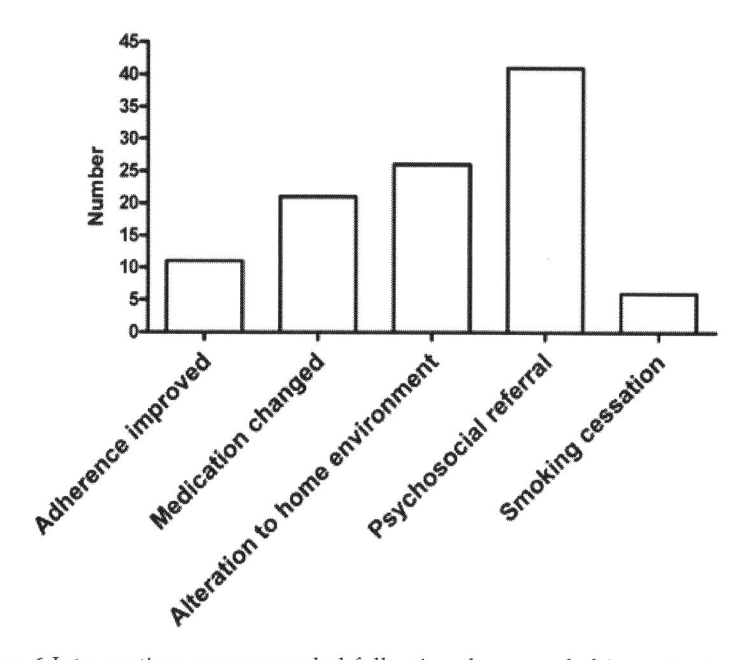

Figure 6 Interventions recommended following the nurse led investigations and home visit in our group of patients.
More than one interventions could be recommended for each child.
Reproduced from: Bracken M., L. Fleming, P. Hall, N. Van Stiphout, C. Bossley, E. Biggart, N. Wilson and A. Bush. 2009. The importance of home visits in children with problematic asthma. Arch Dis Child 94: 780–4 (with permission).

DEFINITIONS AND EXPLANATION OF WORDS AND TERMS

Allergen: A foreign substance that can produce an IgE mediated response. The immune system recognises allergens as 'foreign or dangerous' but cause no reaction for most people.

Adherance: The level and extent to which a patient complies with a prescribed course of medication/treatment.

Beyond guideline treatment: Experimental high risk treatments above recommended doses such as high dose inhaled corticosteroids, prolonged oral steroids, Intramuscular triamcinolone, and steroid sparing agents such as cyclosporin, methotrexate and azathioprine.

Problematic, severe asthma: Persisting symptoms 3 or more days a week, frequent exacerbations (one or more a month) or a previous admission to PICU despite being on high dose inhaled corticosteroids, long acting B2 agonist, leukotriene receptor agonist (or failed trial).

Skin Prick Testing: A test to determine whether a person has sensitivity to a particular allergen. It is performed by putting small drops of specific allergens onto the skin and making a small prick in the skin service using a lancet. A positive reaction is demonstrated by a wheal that can be measured and compared to a positive and negative control. It is a safe, easy and painless test to perform.

Specific IgE test: A blood test to identify the amount of specific immunoglobulin E present to certain allergens.

Urinary Cotinine: A laboratory test using a sample of the patient's urine to determine exposure to environmental tobacco smoke.

LIST OF ABBREVIATIONS

ACT	:	Asthma control test
CAMHS	:	Children's and adolescent mental health services
FeNO	:	Exhaled nitric oxide
FEVI1	:	Forced expiratory volume in the first second
HEPA	:	High Efficiency Particulate Air
HDM	:	House dust mite
ICS	:	Inhaled corticosteroid steroid
IgE	:	Immunoglobulin E
PAL	:	Persistent airflow limitation
pMDI	:	Pressurised metered dose inhaler
SPT	:	Skin prick test

REFERENCES

Ayres, J.G., J.F. Miles and P.J. Barnes. 1998. Brittle asthma. Thorax 53: 315–321.

Barry, C.A., F.A. Stevenson, N. Britton, N. Barber and C. Bradley. 2001. Giving voice to the life world. More humane, more effective medical care? A qualitative study of doctor patient communication in general practice. Soc Sci Med 53: 487–505.

Bossley, C.J., S. Saglani, C. Kavanagh, D.N.R. Payne, N. Wilson, L. Tsartsali, M. Rosenthal, I. Balfour-lynn, A.G. Nicholson and A. Bush. 2009. Corticosteroid responsiveness and clinical characteristics in childhood difficult asthma. Eur Respir J (in press).

Bracken. M., L. Fleming, P. Hall, N. Van Stiphout, C. Bossley, E. Biggart, N. Wilson and A. Bush. 2009. The importance of home visits in children with problematic asthma. Arch Dis Child 94: 780–4.

Bush, A. and S. Saglani. 2010. Management of severe asthma in children. Lancet (in press).

Bush, A., G. Hedlin, K.H. Carlsen, F. de Benedictis, K. Lodrup-Carlsen and N. Wilson. 2008. Severe childhood asthma: a common international approach? Lancet 372: 1019–21.

Chalmers, G.W., K.J. Macleod, S.A. Little, A. Thompson, C. McSharry and N. Thomson. 2002. Influence of cigarette smoking on inhaled corticosteroid treatment in mild asthma. Thorax 57: 226–30.

Dean, A., J. Walters and A. Hall. A systematic review of interventions to enhance medication adherence in children and adolescents with chronic illness. Arch Dis Child Online 2010.

De Groot, E., E. Duivernab and P. Brand. 2010. Co-morbidities of asthma during childhood: possibly important, yet poorly studied. Eur Respir 36: 671–678.

de Jongste, J.C., S. Carraro and W.C. Hop. 2009. CHARISM study group, Baraldi E. Daily telemonitoring of exhaled nitric oxide and symptoms in the treatment of childhood asthma. Am J Respir Crit Care Med 179: 93–7.

Frith, J., L. Fleming, C. Bossley, N. Ullmann and A. Bush. The complexities of defining atopy in severe childhood asthma. *Clinical and experimental Allergy* (in press)

Green, R.H., C.E. Brightling, S. McKenna, B. Hargadon, D. Parker, P. Bradding, A. Wardlow and I.D. Pavord. 2002. Asthma exacerbations and sputum eosinophil counts: a randomised controlled trial. Lancet 360: 1715–21.

Gotzsche P.C. and H.K. Johansen. 2008. House dust mite control measures for asthma: systematic review. Allergy 63: 646–659.

Heaney, L.G., E. Conway, C. Kelly et al. 2005. Prevalence of psychiatric morbidity in a difficult asthma population: relationship to asthma outcome. Respir Med 99: 1152–1159.

Hedlin, G., A. Bush, K. Lodrup Carlsen, F. Wennergren, F. De Benedictis, E. Melen, J. Paton and N. Wilson. 2010. On behalf of the Problematic Severe Asthma in Childhood Initiative group. Problematic severe asthma in children, not one problem but many: a GA2LEN. Initiative. Eur Respir J 36 196–201.

Horner, S.D. 2006. Home visiting for intervention delivery to improve rural family asthma management. J.Community Health Nurse 23: 487–505.

Juniper, E.F., G.H. Guyatt, D.H. Feeny. P.J. Ferrie, L.E. Griffith and M. Townsend. 1996. Measuring quality of life in children with asthma. Qual Life Res 5: 35–46.

Juniper, E.F., G.H. Guyatt, D.H. Feeny, P.J. Ferrie, L.E. Griffith and M. Townsend. 1996. Measuring quality of life in the parents of children with asthma. Qual Life Res 5: 27–34.

Krawiec, M., R. Covar and G. Larsen. 2008. CARE Network. Azithromycin or montelukast as inhaled corticosteroid-sparing agents in moderate-to-severe childhood asthma study. J Allergy Clin Immunol 122: 1138–1144.

Murray, C.S., G. Poletti, T. Kebadze, J. Morris, A. Woodcock, S. Johnston and A. Custovic. 2006. Study of modifiable risk factors for asthma exacerbations:

virus infection and allergen exposure increase the risk of asthma hospital admissions in children. Thorax 61: 376–82.

Ogden, J., K. Bavalia, M. Bull, S. Frankum, C. Goldie, M. Gosslau, A. Jones, S. Kumar and K. Vasant. 2004. "I want more time with my doctor": a quantitative study of time and the consultation. Fam Pract 21: 479–483.

Orrell-Valente, J.K., L.G. Jarlsberg, L.G. Hill and M. Cabana. 2008. At what age do children start taking daily asthma medicines on their own? Paediatrics 122: E1186–E1192.

Richardson, L.P., P. Lozano, J. Russo, E. Mc. Cauley, T. Bush and W. Katon. 2006. Asthma symptom burden: relationship to asthma severity and anxiety and depression symptoms. Paediatrics 118: 1042–1159.

Sandberg, S., J.Y. Paton , S. Ahola, D. McCann, D. McGuinness, D. Hillary, R. Clive and O. Hannu. 2000. The role of acute and chronic stress in asthma attacks in children. Lancet 356: 982–7

Struck, R.C., L.B. Bacharier, B.R. Phillips, S.J. Szefler, R.S. Zeiger, V.M. Chimchilli, F.D. Martinez, R.F. Lemanske jr., L.M. Taussig, D.T. Mauger, W.J. Morgan, C.A. Sorkness, I.M. Paul, T. Guilbert, M. Krawiec, R. Covar and G. Larsen. 2008. CARE Network. Azithromycin or montelukast as inhaled corticosteroid-sparing agents in moderate-to-severe childhood asthma study. J Allergy Clin Immunol 122: 1138–1144.

Sulakvelidze, I., M.D. Inman, T. Rerecich and P.M. O'Byrne. 1998. Increases in airway eosinophils and interukin-5 with minimal bronchoconstriction during repeated low-dose allergen challenge in atopic asthmatics. Eur Respir J 11: 821–7.

Szefler, S.J., H. Mitchell, C.A. Sorkness, P.J. Gergen , G.T. O'Connor, W.J. Morgan, M. Kattan, J.A. Pongracic , S.J. Teach, G.R. Bloomberg, P.A. Eggleston, R.S. Grunchalla, C.M. Kercsmar, A.H. Liu, J.J. Wildfire, M.D. Curry and W.W. Busse. 2008. Management of asthma based on exhaled nitric oxide in addition to guideline-based treatment for inner-city adolescents and young adults: a randomised controlled trial. Lancet 372: 1065–72.

Watson, J.P., P. Cowen and R.A. Lewis. 1996. The relationship between asthma admission rates, routes of admissions, and social deprivation. Eur Respir J 9: 2087–2093.

Wood, R.A., M.D. Chapman, N.F Adkinson Jr. and P.A. Eggleston. 1989. The effects of cat removal on Fel d I content in household dust samples. J Allergy Clin Immunol 83: 730–4.

10

Asthma Therapy in Elderly Patients

Andrzej Bozek[1,a,*] and Jerzy Jarzab[1,b]

ABSTRACT

Asthma is a common disease in the elderly. Changes that occur in the aging process affect diagnosis and treatment of this disease. Under-diagnosis and under-treatment are typical in older people. Risk factors and comorbidities, especially heart failure, impaired cognition, asthma and chronic obstructive pulmonary disease, overlap and should be analysed carefully before initiating anti-asthmatic therapy. Specific challenges encountered when treating elderly people include worse adherence to anti-asthmatic therapy and increased risk of adverse responses to anti-asthmatic drugs. However, the main goals for successful management of asthma are similar to those in young patients and include maintaining control of asthma symptoms, excluding nocturnal awakenings due to asthma, reducing medication use, preventing asthma exacerbations, improving pulmonary function and decreasing asthma-induced mortality. Inhaled corticosteroids are the most effective anti-asthmatic drug in asthma, but poor adherence to their use is still observed in elderly people. Good inhaler technique, adherence to treatment and participation in educational training sessions improve outcomes.

[1]Clinical Department of Internal Medicine, Dermatology and Allergology, Medical University of Silesia, Zabrze, M. Sklodowskiej-Curie 10, 41-800 Zabrze, Poland;
Email: andrzejbozek@o2.pl
Email: jerzy.jarzab@gmail.com
*Corresponding author

List of abbreviations after the text.

INTRODUCTION

Asthma in the elderly population is a globally increasing problem. The prevalence of asthma in the elderly ranges from 6–17% (1,2,3). It is due to an aging population, a high prevalence of allergies, smoking, exposure to pollution and other factors (2,3,4). Older patients with asthma are divided into two groups: those with long-standing asthma from childhood and those with late onset in sixth decade of life (aged 60–65). Patients with long-standing asthma have shorter symptom-free periods, more emergency interventions, more hospitalisations and worse spirometric values than patients with asthma onset after 60–65 years old (5). Elderly asthmatic patients are often under-diagnosed and consequently mistaken for chronic obstructive pulmonary disorder (COPD) (1,6,7). Under-treatment of asthma is common in older people. Many studies have shown a difference between the recommended guidelines for asthma and actual therapy, and reveal poor adherence to anti-asthmatic drugs, especially inhaled corticosteroids (3,5,6). Uncontrolled disease, poor quality of life and mortality rate in asthma may be related to insufficient anti-asthmatic therapy in aged patients.

PRACTICE AND PROCEDURES

Treatment of asthma in older patients, as in the younger population, is based on world guidelines. GINA (Global Initiative for Asthma), ATS (American Thoracic Society), AAAI (American Academy of Allergy, Asthma and Immunology) and other groups focus on therapy for asthma in children and adults. There are little data regarding patients aged >60. The needs of older people with asthma are partly similar to those of young asthmatics, but the psychosocial problems and safety profiles of treatment options worsen with age (8,9).

The specific treatment of elderly people with asthma concerned the following: worse adherence to anti-asthmatic therapy, increased risk of anti-asthmatic drug adverse events especially after $beta_2$-agonists and oral corticosteroids and high risk of mortality, which particularly apply to older women simultaneously suffering from heart failure, diabetes and impaired cognition (1,9).

Lack of patient understanding and management of asthma is a consequence of poor communication between patients and medical staff (doctors, nurses etc). It leads to poor adherence and unwillingness to continue the asthma therapy. Patients, their guardians and medical staff have different perspectives on the disease and the aims of treatment. Lack of patient education contributes to different fears, unrealistic expectations and other problems (1,3,4,10,11). The mode of education used with elderly asthmatics may depend on patient preferences (information leaflets, booklets, videos), but should be preceded by education by medical staff. It should be performed at every patient visit, and it requires sufficient time (1,4).

According to all of the asthma guidelines, medications used to treat asthma are divided into two main groups: rescue medications and controllers (4). The first group is represented by the inhaled short-acting beta$_2$-agonist (SABA) and inhaled ipratriopium bromide. The second group includes anti-inflammatory medications such as inhaled glucocorticosteroids (ICSs), which are the staple drug of asthma treatment, leukotriene-modifying agents (LTRAs), long-acting inhaled beta$_2$-agonists (LABA), sustained release theophylline and sodium cromoglycate. However, particularly in severe asthma, the new groups of drugs are recommended: methotrexate, omalizumab (anti IgE therapy) or phosphodiesterase (4). These are also used in the elderly.

SABAs effectively induce bronchodilation during asthma attacks. The guidelines recommend SABA as required. The optimal tolerated dose is strictly individual in elderly patients. However, they frequently cause overdoses and serious side effects (Table 3). Additionally, hypoxia is a risk after too frequent use by elderly asthmatics, particularly during asthma exacerbations when older patients are unable to recognise the nature of their dyspnoea (1,4,5,9).

LABAs are the most effective drugs and should be always associated with inhaled corticosteroids. Their synergistic effect is very useful in therapy of chronic moderate or severe asthma. However, increasing bronchial hyperreactivity and worse asthma exacerbations requiring hospitalisation are observed in elderly patients who use LABA. In general, older patients should start on inhaled corticosteroids before electing to add LABA if required (1,5,9).

Inhaled ipratropium a less effective bronchodilator, but it is more efficient in elderly people. The cholinergic tonus and low response to beta agonists probably influence this choice. These drugs are safer and represent a good alternative in patients with side effects after SABA administration. Anticholinergics have a slower onset of action. The role of long-term anticholinergics such as tiotropium bromide is well defined in COPD but is still under investigation in asthma (1,4,5,9).

Inhaled corticosteroids are the most effective controller medications due to their wide spectrum of anti-inflammatory actions. Several studies have demonstrated that treatment with inhaled corticosteroids for 1 month significantly reduces pathological signs of airway inflammation and airway hyperresponsiveness, leading to better control of asthma (3,4,5,9,12,13). The goal of treatment with inhaled corticosteroids is to avoid the need for systemic corticosteroids, though this is not always possible in older asthmatics, partly due to specific concerns. Poor compliance with inhaled corticosteroid use and fear of side effects are common. In clinical practice, under-treatment with this type of drug is common (3,14). Drug delivery by the inhaled route using corticosteroids offers the best efficacy ratio for asthma therapy, but ineffective inhalation techniques remains a problem limits control of asthma (5,9).

There are many factors that impair inhaler technique: manual problems, weak vision, impaired cognitive function, unfamiliarity with the principles of using MDI inhalators and decreased inspiratory flow, which varies from 40–60% (5,9). Elderly asthmatic patients should be carefully instructed in self-management. The inhaler technique should be demonstrated at every medical visit by the doctor and medical staff. However, some patients will be unable to perform it. A dry powder inhaler might be a better option, but about 40% patients have difficulty performing ineffective inhalation even with dry powder (1,5,6,14).

The compact combination LABA and ICS inhalers are commonly used by elderly asthmatics. Better adherence to use of LABA with ICS on demand is more effective and better prevents asthma exacerbations in elderly patients. However, these combination inhalers are not always necessary, particularly in patients in whom the adverse effects of both drugs can be substantial (1).

Leukotriene-modifying agents are useful inflammatory mediators for anti-asthmatic treatment. These drugs improve lung function, reduce exacerbation frequency and improve symptoms such as exercise-induced dyspnoea. They are efficacious as single agents, but asthma control improves when inhaled steroids are added.

Leukotriene-modifying agent administration is simple (pills once or twice per day) and thus encourage good adherence in elderly asthmatics (1,5,9).

Sustained-release preparations of oral theophyllines can be used in poorly controlled asthma and when other anti-asthmatic drugs are insufficient. Theophyllines have bronchodilatory and anti-inflammatory effects. It may be an alternative to LABA as a supplement to inhaled corticosteroids (1). Adverse effects of anti-asthmatic drugs are summarised in Table 1.

Table 1 Adverse effects after anti-asthmatic drugs.

Category	**side effects**
beta$_2$-adrenoreceptor agonists	increase risk of mortality in elderly patients with cardiovascular diseases: tachycardia, tachyarrhythmia, tremor, nervous tension, headache, asthma exacerbations especially with monotherapy
anticholinergic agents	good safety profile in general urinary retention, increased intraocular pressure, tachycardia, xerostomia
Corticosteroids	
inhaled	good safety profile in general but may cause dysphonia, throat irritation, oropharyngeal thrush, candidiasis
oral	osteoporosis, diabetes mellitus, glaucoma, arterial hypertension, suppression of the hypothalamic-pituitary-adrenal axis, myopathy, cataracts fractures, immunoparesis, loss of attention span and memory, easy bruising-especially in elderly (elderly patients with high dose of inhaled corticosteroids or oral steroids should take supplements of calcium and vitamin D)
sustained-release oral theophyllines	headaches, insomnia, gastrointestinal disturbances, arrhythmias, hypokalemia
leukotriene-modifying agents	good safety profile in general, very rarely hepatitis

unpublished material of author.

A stepwise approach to the treatment of asthma is recommended. According to GINA recommendations, patients are assigned to the treatment step most appropriate to the severity of their asthma, the aim being to achieve early symptom control before titrating treatment for maintenance of control (4). However, in older patients there are some modifications. The first category is intermittent asthma, in which patients need only SABA as required. In older patients, ipratropium in MDI offers a better safety profile (1,3,5,9). In mild chronic asthma, the inhaled corticosteroids must be started at a dose appropriate to the degree needed to control asthma. However, in this step, leukotriene-modifying agents can be added as monotherapy or in combination with inhaled corticosteroids. In moderate asthma, a LABA could be added cautiously due to the risk of cardiovascular side effects. Alternatives include increased doses of inhaled steroids, leukotriene-modifying agents (if not previously used) and sustained-release (SR) theophylline. SR theophylline can be used in patients with low adherence. In patients with severe asthma, high-dose inhaled corticosteroids should be administered with drugs used in lower steps. However, oral steroids are frequently necessary (1,5,9).

GINA distinguishes uncontrolled, partly-controlled and well-controlled asthma and recommends appropriate treatment for controlling asthma symptoms for each category (4).

The main goals for successful management of asthma are as follows: maintain control of asthma symptoms, including normal activities with effort; exclude nocturnal awakenings due to asthma; reduce drug prescriptions as possible; prevent asthma exacerbations; improve pulmonary function as close to normal levels as possible; and prevent asthma mortality (4). The self-management plan is necessary and influences adherence and therapeutic results (15,16).

In elderly patients, it is very important to avoid the adverse effects of asthma medications. In the elderly, asthma is comorbid with other chronic disorders such as heart failure, diabetes, arterial hypertension and dementia. This leads to use of many drugs with interactions and adverse effects (1,3,17). Polypharmacy is a significant problem in older people. It is responsible for many geriatric problems as vertigo, orthostatic hypotony, mental disorders and others. Additionally, asthma-specific guidelines do not address the preferences of individual patients. COPD and asthma can be difficult to distinguish but often coexist in older patients. This introduces challenges to proper therapy because COPD treatment is based primarily on bronchodilators such as beta agonists and anticholinergics but lacks the maintenance therapy of asthma controller drugs (1,3,5).

Adherence can be a complex issue in elderly people with bronchial asthma and other chronic diseases. Studies have shown that approximately 50% of asthmatic patients on long-term therapy fail to take the medications as directed at least part of the time. Non-adherence may be defined in a nonjudgmental way as the failure to take treatment as agreed upon by the patient and health care professional (15,18).

Depressive symptoms, as a common problem in elderly patients, also corresponded with adherence to anti-asthmatic treatment (3,8,10,15,19). It is difficult to control adherence to therapy in older people. Commonly used questionnaires to control the use of drugs are insufficient (3,15); assistance is needed from medical staff, guardians and family. The estimation of drug adherence using electronic diaries and drug package assessment is more reliable than other methods. Importantly, the treatment plan should be carefully analysed by the doctor, the patient and the patient's guardians before starting anti-asthmatic therapy.

Physical training programs have been designed for asthmatic subjects with the goal of improving physical fitness, neuromuscular coordination and self-confidence (20,21). However, these programs are still unappreciated, especially in older patients. Habitual physical activity increases physical fitness and lowers ventilation during mild and moderate exercises that are correctly adjusted for the elderly patient. It reduces the likelihood of provoking exercise-induced asthma. Exercise training may also reduce the perception of breathlessness. Moreover, many patients report they are symptomatically better when fit, though results from trials have varied. In people with asthma, physical training can improve cardiopulmonary fitness without changing lung function. There are also observations that decreased depressive symptoms and increased quality of life can be expected during regular training in older patients (15,17).

There are no established physical training programs for elderly asthmatic patients. Regular physical training should be undertaken for 15–20 minutes two to three times a week. Swimming has been shown to improve overall well-being. Other diseases that handicap or decrease daily activities could reduce the intensity and time of trainings (3,20,21).

Obesity is a known risk factor for variable airflow limitation. Obesity makes stretching of the airways difficult, although evidence that it worsens asthma is limited. On the other hand, it is important to note that low weight in elderly patients increases the risk of geriatric problems including mortality (3,4,12,22).

Smoking is a significant problem in older asthmatic patients. It can induce impaired corticosteroid response, decreased lung function and increased mortality. In older patients, a positive history of tobacco smoking can contribute to misdiagnosis of COPD and result in inappropriate therapy. About 50% of older patients have a positive history of smoking. The health-related beliefs of older smokers can differ from those of younger smokers, and these differences might affect how they choose to stop smoking. However, physicians might be less aggressive with counselling for smoking cessation in patients aged > 60 (3,23).

Asthma Exacerbations

In elderly patients, asthma exacerbations are still common and death is more likely than in young asthmatics (1,4,22). The factors that could induce asthma exacerbations include the lack of asthma diagnosis, undertreated asthma (beta 2 mimetic overdose, low use of inhaled corticosteroids), poor adherence to therapy, inappropriate evaluation of the degree of dyspnoea, low socio-economical status, infections, black race, female sex, asthma severity, impaired cognition and depression symptoms.

Nebulised beta mimetics, oxygen therapy and steroids are basic therapy. Nebulised anticholinergics are also useful. However, other drugs are sometimes needed (4,9).

The treatment of severe asthma treatment should be provided in the emergency unit.

Additional supportive therapy is also important, especially in elderly patients. These include influenza vaccination, pneumococcal vaccination, physical activity as mentioned above and proper therapy of comorbidities such as allergic rhinosinusitis, reflux, heart failure and impaired cognition (3,4,12,13).

Older patients with bronchial asthma require special attention. The role of the nursing staff in teaching good inhaler technique, and in the control of adherence to treatment and participation in education programs improves outcomes (24).

KEY FACTS OF ASTHMA IN THE ELDERLY

- Asthma is a chronic inflammatory disease with recurrent episodes of wheezing, breathlessness, cough and chest tightness. Symptoms are associated with variable airflow limitations.
- Asthma in the elderly is a common disease, and its clinical features are similar to those observed in the young.
- Bronchial asthma and chronic obstructive pulmonary disorder are frequently confused.
- A diagnosis is, at a minimum, based on positive history (breathlessness) and spirometric reversibility.
- Older patients with asthma are divided into two categories: "long-standing" asthma diagnosed in childhood and "late-onset asthma" that started after 30 years of age.
- Nonatopic asthma is more common in older patients; however, atopic asthma is frequent.
- The mortality of asthma is low but significantly rises during exacerbations that require the hospitalisation of older patients.
- The quality of life in elderly patients with asthma is low and is related to depressive symptoms, as is typical in chronic disorders.
- Regular inhaled corticosteroids are the mainstay of the treatment of asthma.

SUMMARY POINTS

- Asthma in elderly patients is commonly misdiagnosed.
- Asthmatic, older patients are usually under-treated and under-diagnosed.
- A stepwise approach to the treatment of asthma using controller and rescue drugs is recommended.

- We recommend good inhaler techniques, an asthma self-management plan, and the minimisation of adverse drug effects after assessing the effectiveness of therapy.
- Physical activity and cessation of smoking are of significant importance in the therapy of asthma.

DEFINITIONS AND EXPLANATION OF WORDS AND TERMS

Asthma: a disorder of variable expiratory airflow obstruction that arises in association with episodic symptoms of wheeze, cough, and dyspnoea. Bronchial inflammation and modifications of the structural elements of the airway wall (remodelling) are pathological components of asthma.

Atopy: the genetic tendency to develop the classic allergic diseases —atopic dermatitis, allergic rhinitis, and asthma. Atopy involves the capacity to produce IgE in response to common environmental proteins such as dust mite, grass pollen, and food allergens.

Older patients: persons aged 64 years and older. There are early elderly patients aged 64–75 years and older elderly aged 75 years and older.

Peak expiratory flow (PEF): the maximum airflow during a forced expiration beginning with the lungs fully inflated. The PEF is reduced in proportion to the severity of the airway obstruction in asthma and COPD (chronic obstructive pulmonary disease).

Dry powder inhaler (DPI): A dry powder inhaler, or DPI for short, is a type of inhaler device for administering asthma medicine. It is similar to a metered dose inhaler, but the medication is in powder form, rather than in liquid form.

Metered dose inhaler (MDI): A metered dose inhaler is a metal and plastic device that people who have asthma can use to get a specific amount of asthma medicine into their airways. When patient pushes on the top of the canister, it releases a metered dose of medicine into plastic tube for the patient to inhale. MDI delivered through spacers (holding chambers) is more effective and is recommended.

Nebuliser/nebulisations: These describe another inhalation system. A nebuliser is used to transform liquid anti-asthmatic medication into vapour for use. Pressured air is pumped through the liquid to form a fine mist which can be inhaled through a mask or mouthpiece.

Spirometry: This is a pulmonary lung function test that measures the amount of air inhaled (volume) and exhaled as a function of time. During a spirometry test, the patient places his/her mouth over the mouthpiece of the spirometer, breathes in deeply, and then exhales as forcefully as possible. Spirometry gives health professionals two important indicators of disordered lung function: FVC (forced vital capacity—how much air can be blown out of the lungs) and FEV1 (forced expiratory volume—the amount of air the patient can blow out in 1 second). It may used to diagnose asthma.

COPD: Chronic obstructive pulmonary disease is a group of lung function diseases that cause obstruction of the airways. This results in a decreased ability to move air in and out of the lungs. Diseases classified as COPD include emphysema, chronic bronchitis and bronchiectasis. Even with treatment, COPD is not completely reversible and usually worsens over time.

Controllers: anti-inflammatory, anti-asthmatic drugs that help control the underlying inflammation in the airways associated with asthma. There are mainly inhaled corticosteroids or leukotriene-modifying agents. Those agents are taken daily independent of the occurrence of asthma symptoms.

Beta-agonists: agents (drugs) that can stimulate beta-adrenergic receptors and protect against bronchospasm.

Rescue medications: drugs that provide relief from asthma symptoms and that are the most commonly used asthma medications. Relievers do not reduce the underlying inflammation associated with asthma and thus do not prevent asthma attacks.

LIST OF ABBREVIATIONS

COPD	:	chronic obstructive pulmonary disease
DPI	:	dry powder inhaler
FEV1	:	forced expiratory volume
GINA	:	Global Initiative for Asthma
ICSs	:	inhaled glucocorticosteroids
IgE	:	immunoglobulin E
LABA	:	long-acting inhaled $beta_2$-agonists

LTRAs	:	leukotriene-modifying agents
MDI	:	metered dose inhaler
PEF	:	Peak expiratory flow
SABA	:	short-acting beta$_2$-agonist

REFERENCES

Barua, P. and M.S. O` Mahony. 2005. Overcoming gaps in the management of asthma in older patients. New insight. Drugs Aging 22: 1029–1059.

Burrows, B., R.A. Barbee, C.M Cline, R.J. Knudson and M.D. Lebowitz. 1991. Characteristics of asthma among elderly adults in a sample of the general population. Chest 100: 935–942.

Gibson G.B., V.M. McDonald and G.B. Marks. 2010. Asthma in older patients. Lancet 376: 803–813.

Global Initiative for Asthma. Gina Report. Global Strategy for Asthma Management and Prevention 2008. Retrieved May 20, 2008 from http://www.gina.org/

Chotirmall, S.H., M. Watts, P. Branagan, C.F. Donegan, A. Moore and N.G. Mc Elvaney. 2009. Diagnosis and management of asthma in older patients. J Am Geriatr Soc 57: 901–909.

Banerjee, D.K., G.S. Lee, S.K. Mailk and S. Dalz. 1987. Underdiagnosis of asthma in the elderly. Br. J Dis Chest 81:23–29.

van Schayck, C.P., F.M. van der Heijden, G. van dem Boom, P.R. Tirimanna and C.L. Herwaarden. 2000. Underdiagnosis of asthma: Is the doctor or patient to blame? DIMCA project. Thorax 55: 562–565.

Cousens, N.E., D.P. Goeman, J.A. Douglass and C.R. Jenkis. 2007 The needs of older people with asthma. Aust Fam Physician 36: 729–731.

Urso, D.L. 2009. Asthma in the elderly. Curr Gerontol Geriatr Res 858415 Epub 2009 Oct 27.

Byles, J.E. 2005. How do the psychosocial consequences of ageing affect asthma management? Med. J. Aust. 183 (1 Suppl): S30–2.

Navaratnam, P., S.S. Jayawant, C.A. Pedersen and R. Balkrishhnan. 2008. Asthma pharmacotherapy prescribing in the ambulatory population of the United States: evidence of nonadherence to national guidelines and implications for elderly people. J Am Geriatr Soc 56: 1312–1317.

King, M.J. and N.A. Hanania. 2010. Asthma in the elderly: current knowledge and future directions. Curr Opin Pulm Med 16: 55–59.

de Vries, U. and F. Petermann. 2008. Asthma among senior adults. Internist (Berl) 49: 1340–1341.

Wieshammer, S. and J. Dreyhaupt. 2008. Dry powder inhalers: which factors determine the frequency of handling errors? Respiration 75: 18–25.

Bozek, A. and J. Jarzab. 2010. Adherence to asthma therapy in elderly patients. J Asthma 47: 162–165.

Rank, M.A., G.W. Volcheck, T.J. Li, A.M. Patel and K.G. Lim. 2008. Formulating an effective and efficient written asthma action plan. Mayo Clin Proc 83: 1263–1270.

Gibson, P.G., P.I. Talbot and R.C. Toneguzzi. 1995. Self—management, autonomy, and quality of life in asthma. Chest 107: 1003–1008.

Gupta M. and M.S. O`Mahony. 2008. Potential adverse effects of bronchodilators in the treatment of airways obstruction in older people: recommendation for prescribing. Drugs Aging 25: 415–443.

Smith, A., J.A. Krishnan, A. Bilderback, K.A. Riekert, C.S. Rand and S.J. Bartlett. 2006. Depressive symptoms and adherence to asthma therapy after hospital discharge. Chest 130: 1034–1038.

Ram, F.S., S.M. Robinson, P.N. Black and J. Picot. 2005. Physical training for asthma. Cochrane Database Syst Rev 19:CD001116.

Ritz, T., D. Rosenfield and A. Steptoe. 2010. Physical activity, lung function, and shortness of breath in the daily life of individuals with asthma. Chest 138: 913–918.

Bellia, V., C. Pedone, F. Catalano, A. Yito, E. Davi, S. Palange, R. Forastiere and R.S. Incalzi. 2007. Asthma in the elderly. Morality rate and associated risk factors for morality. Chest 132: 1175–1182.

Brown, D.F., J.B. Croft, A.P. Schenck, A.M. Malarcher, W.H. Giles and R.J. Simpson Jr. 2004. Inpatient smoking—cessation counseling and allcause morality among the elderly. Am J Prev Med 26: 112–118.

Wilhelmsson, S. and M. Lindberg. 2007. Prevention and health promotion and evidence—based fields of nursing—a literature review. Int J Nurs Pract 13: 254–265.

11

Treatment of Pediatric Asthma

***Kenny Yat-Choi Kwong*[1,*] and *Peter Huynh*[2]**

ABSTRACT

Asthma is a leading cause of morbidity and health burden in children worldwide. Diagnosis of asthma is based upon a combination of clinical symptoms and lung function. Goal of asthma treatment in children is to achieve asthma control. This results in reduction of symptoms and risk of exacerbations. Asthma control is measured with two domains: impairment and risk. Impairment is a "snapshot" of present clinical symptoms and risk is an assessment of potential future risk of asthma exacerbation. Measurement of impairment is based upon a combination of present clinical symptoms and lung function. Asthma control is usually attained by use of long term anti-inflammatory medications including inhaled corticosteroids, long acting beta agonists and leukotriene receptor antagonists. Inhaled corticosteroids are first line agents for persistent asthmatics. For patients who are uncontrolled on inhaled corticosteroids, the addition of second controllers improves asthma control. Short term relief of asthma symptoms is usually achieved with the use of short acting beta agonists. Treatment of co-morbid conditions and adherence to compliance also improves asthma control. Asthma controller medications are stepped up when asthma control is poor and stepped down when asthma control is sustained. Finally asthma self management plans help promote asthma control.

[1]Division of Allergy-Immunology, Department of Pediatrics, 1000 West Carson Street, N-25 Box 491, Torrance California 90808, USA; Email: kkwongusc@yahoo.com
[2]Division of Allergy-Immunology, Department of Pediatrics, LAC+USC Medical Center, 1801 East Marengo Street Room 1G1, Los Angeles California 90033, USA; Email: phuynhmd@sbcglobal.net
*Corresponding author

List of abbreviations after the text.

INTRODUCTION

Uncontrolled asthma is a leading cause of morbidity in children worldwide, incurring high socio-economic costs (EPR-3 2007, GINA 2002). Despite the development of new asthma medications, asthma control in children remains poor in the 21st century. This chapter focuses on the goals of care in pediatric asthma and strategies to achieve asthma control. Treatment of acute symptoms and exacerbations will also be briefly discussed.

DEFINITION AND DIAGNOSIS

Asthma is a chronic disease characterized by diffuse airway inflammation, episodic airway obstruction that is at least partially reversible, and airway hyperresponsiveness. Coughing, shortness of breath, chest tightness and wheezing are cardinal clinical symptoms associated with asthma. There is no single confirmatory test for asthma and diagnosis of asthma relies upon careful review of symptoms in combination with physical examination and objective measures of lung function (Table 1). Spirometry is the recommended method of objectively measuring airflow obstruction and reversibility. Diagnosis of asthma is supported by reversible airflow obstruction as measured by forced expiratory volume (FEV1) by more than 12% after bronchodilator therapy (ATS 1995). Alternative diagnoses must be excluded (Table 2).

Table 1 Diagnosis of Asthma.

Diagnosis of asthma requires the following to be satisfied:
Symptoms associated with episodic airflow obstruction and/or airway hyperresponsiveness
Presence of reversible airway obstruction
Other diagnosis are excluded
Methods used to make diagnosis of asthma:
Detailed history
Physical examination
Spirometry to demonstrate airway obstruction and reversibility
Other studies to exclude other diagnosis

Table 2 Differential diagnosis of asthma in children.

Differential diagnosis
Upper airway pathology
Allergic rhinitis
Sinusitis
Large airway obstructions
Foreign bodies
Vocal cord dysfunction
Vascular rings and laryngeal webs
Laryngotracheomalacia, tracheal stenosis or bronchostenosis
Lymphadenopathy or tumors
Small Airway Obstructions
Viral bronchiolitis and bronchiolitis obliterans
Cystic fibrosis
Bronchopulmonary dysplasia
Cardiac disease
Others
Chronic cough not due to asthma
Gastroesophageal reflux or aspiration

Asthma diagnosis in younger children is often challenging as episodic wheezing and cough are also common in children without asthma (Huynh et al. 2010). Many children have been reported to wheeze in their first year of life. Diagnosis is further hampered by difficulty in obtaining objective measurements of lung function in children less than 5 years. Moreover, there are different phenotypes of wheezing disease in children, and virally induced wheezing in non-atopic children is often transient. These patients should not be prescribed prolonged asthma therapy. Alternatively, treatment must not be delayed in children whose asthma is likely to persist onto school age.

Atopic predisposition in children is an important risk factor associated with the development of asthma. The Asthma Predictive Index (API) is an instrument that helps identify children who are likely to have persistent asthma (Castro-Rodriquez et al. 2000). This is a clinical index based on the presence of wheeze before the age of 3 years. Presence of one major risk factor or two of three minor risk factors has been shown to predict the presence of asthma in later childhood with positive and negative predictive values of 76% and 95% respectively (Table 3).

Table 3 Asthma Predictive Index.

Wheezing criteria: For children younger than 3 years of age with 4 or more episodes of wheezing during the previous year Plus		
1 Major	**or**	**2 Minor**
Family history of asthma		Wheezing apart from colds
Atopic dermatitis		Peripheral eosinophilia > 4%
		Allergic rhinitis
4 episodes of wheezing plus 1 major or 2 minor criteria has a positive predictive index of 76% to have asthma into school age. Negative predictive index is 95%		
Adapted from Castro-Rodriguez et al. Am J Respir Crit Care Med 2000: 162.		

GOALS OF LONG TERM ASTHMA TREATMENT

Goals of asthma treatment include control of disease and symptoms; reduce/eliminate risk of exacerbations; and modify disease progression with minimal side effects from therapeutic agents. International guidelines have strongly recommended "asthma control" as a goal of long term asthma treatment (EPR-3 2007, GINA 2002). This results in reduction of symptoms, improved quality of life, and importantly, reduction in the risk of exacerbations. A high degree of asthma control is likely to significantly lower the risk of asthma exacerbations and emergency department visits/hospitalizations in asthmatic children (Kwong et al. 2008) (Fig. 1).

ASTHMA CONTROL

United States asthma guidelines measure asthma control using two domains: impairment and risk (EPR-3 2007). Impairment is a "snapshot" of the patient's clinical symptoms and lung function at a single time point. Risk is an assessment of the future risk of asthma related morbidity; chiefly, the likelihood of an asthma exacerbation but also risk from medications and decline in lung function. Measurement of impairment is based upon the frequency of present symptoms and lung function, while the risk of a future asthma attack is based upon the number of previous exacerbations requiring the need for systemic steroids (Fig. 2). Global asthma guidelines (GINA 2002) do not delineate impairment and risk, but these measures are incorporated into assessment of asthma control (Fig. 3).

Year 1 asthma control and time elapsed with no acute exacerbations during year 2 in patients aged 3 to 5 years (A), 6 to 11 years (B), and 12 to18 years (C).

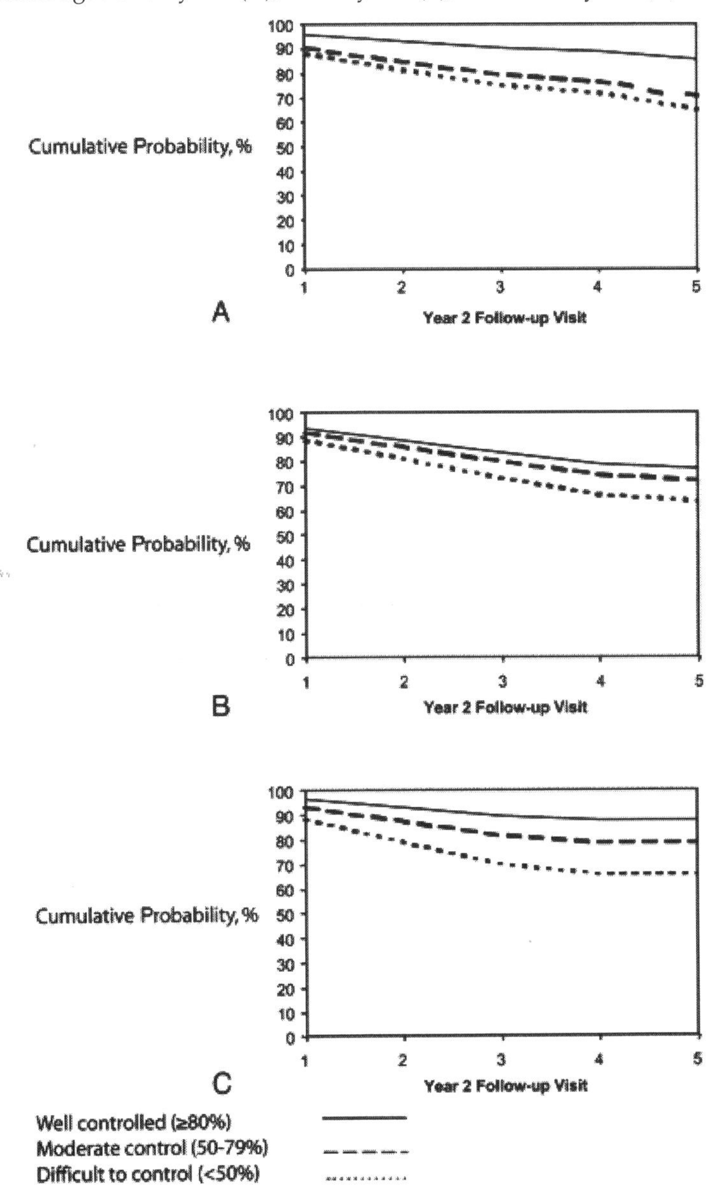

Kwong et al. Ann allergy asthma immunol 2008: 101: 144–152.

Figure 1 Degree of asthma control measured over 1 year is aasociated with reduced risk of asthma exacerbation during the subsequent year in inner city asthmatic children.

FIGURE 2A. ASSESSING ASTHMA CONTROL IN CHILDREN 0-4 YEARS OF AGE.

Components of Control		Classification of Asthma Control (0-4 years of age)		
		Well Controlled	**Not Well Controlled**	**Very Poorly Controlled**
Impairment	Symptoms	≤2 days/week	>2 days/week	Throughout the day
	Nighttime awakenings	≤1x/month	>1x/month	>1x/week
	Interference with normal activity	None	Some limitation	Extremely limited
	Short-acting beta$_2$-agonist use for symptom control (not prevention of EIB)	≤2 days/week	>2 days/week	Several times per day
Risk	Exacerbations requiring oral systemic corticosteroids	0-1 per year	2-3 per year	>3 per year
	Treatment-related adverse effects	Medication side effects can vary in intensity from none to very troublesome and worrisome. The level of intensity does not correlate to specific levels of control but should be considered in the overall assessment of risk.		

FIGURE 2B. ASSESSING ASTHMA CONTROL IN CHILDREN 5-11 YEARS OF AGE.

Components of Control				
		Well Controlled	**Not Well Controlled**	**Very Poorly Controlled**
Impairment	Symptoms	≤2 days/week but not more than once on each day	>2 days/week or multiple times on ≤2 days/week	Throughout the day
	Nighttime awakenings	≤1x/month	≥2x/month	≥2x/week
	Interference with normal activity	None	Some limitation	Extremely limited
	Short-acting beta$_2$-agonist use for symptom control (not prevention of EIB)	≤2 days/week	>2 days/week	Several times per day
	Lung function • FEV_1 or peak flow • FEV_1 / FVC	>80% predicted/personal best >80%	60-80% predicted/personal best 75-80%	<60% predicted/personal best <75% predicted
Risk	Exacerbations requiring oral systemic corticosteroids	0-1 per year	≥2 per year	
		Consider severity and interval since last exacerbation		
	Reduction in lung growth	Evaluation requires long-term followup.		
	Treatment-related adverse effects	Medication side effects can vary in intensity from none to very troublesome and worrisome. The level of intensity does not correlate to specific levels of control but should be considered in the overall assessment of risk.		

FIGURE 2C. ASSESSING ASTHMA CONTROL IN CHILDREN ≥ 12 YEARS OF AGE.

Components of Control		Classification of Asthma Control (≥12 years of age)		
		Well Controlled	Not Well Controlled	Very Poorly Controlled
Impairment	Symptoms	≤2 days/week	>2 days/week	Throughout the day
	Nighttime awakenings	≤2x/month	1-3x/week	≥4/week
	Interference with normal activity	None	Some limitation	Extremely limited
	Short-acting beta$_2$-agonist use for symptom control (not prevention of EIB)	≤2 days/week	>2 days/week	Several times per day
	FEV$_1$ or peak flow	>80% predicted/personal best	60-80% predicted/ personal best	<60% predicted/ personal best
	Validated questionnaires			
	ATAQ	0	1-2	3-4
	ACQ	≤0.75	≥1.5	N/A
	ACT	≥20	16-19	≤15
Risk	Exacerbations requiring oral systemic corticosteroids	0-1 per year	≥2 per year	
		Consider severity and interval since last exacerbation		
	Progressive loss of lung function	Evaluation requires long-term followup care		
	Treatment-related adverse effects	Medication side effects can vary in intensity from none to very troublesome and worrisome. The level of intensity does not correlate to specific levels of control but should be considered in the overall assessment of risk.		

Figure 2 U.S. NHLBI Asthma Guidelines Classification of Asthma Control Criteria.
NHLBI 2007 Guidelines for the diagnosis and management of asthma Expert Panel Report 3. Public domain document.

Prior asthma guidelines recommended classification of asthma severity based on initial assessment before controller treatment to determine the baseline "controller naïve" asthma severity in patients. Updated U.S. guidelines still recommend baseline classification of severity but only as an initial reference point. Many asthma patients initially assessed are not controller naïve and asthma activity varies over time often deviating from initial baseline severity. For instance, some children are relatively free of symptoms during the summer months and are classified as intermittent asthmatics at that time; however, they may experience severe symptoms during periods of viral infections. Thus, current guidelines advocate real time assessment of asthma control at every provider encounter. Therapy with a stepwise approach is recommended with "step up" of controller agents when control is poor and "step down" of medications when control is maintained for periods of time.

Characteristic	Controlled (all of the following)	Partly Cotrolled (any measure present in any week)	Uncontrolled
Daytime Symptoms	None (twice or less/week)	More than twice/week	Three or more features of partly controlled asthma present in any week
Limitations of Activities	None	Any	
Nocturnal symptoms/ awakenings	None	Any	
Need for reliever/rescue treatment	None (twice or less/week)	More than twice/week	
Lung function (PEF or FEV1) ‡	Normal	<80% of predicted or personal best (if known)	
Exacerbations	None	One or more/year *	One in any week +

* Any exacerbation should prompt review of maintenance treatment to ensure that it is adequate
+ By definition, an exacerbation in any week makes that an uncontrolled asthma week
‡ Lung function is not a reliable test for children 5 years and younger

Figure 3 GINA Guidelines Assessment of Asthma Control. Global Initiative for Asthma 2006. Public domain document.

COMPLIANCE

Compliance plays a major role in achieving sustained asthma control in asthmatic children. We have shown that compliance is an independent variable in maintaining asthma control in inner city children with asthma once initial disease control is attained (Jones et al. 2007). Poor compliance to medication is seen in all severities of asthma, and contributes to poor asthma control, increased morbidity, and mortality (Gamble et al. 2009). Poor patient understanding of the disease, difficulty of health care access, socioeconomic issues and complexity of treatment have been cited as reasons for sub-optimal compliance (Howell et al. 2008).

TREATMENT OF CO-MORBID CONDITIONS

Upper airway disease including allergic rhinitis/sinusitis and gastro-esophageal reflux (GERD) contribute towards asthma morbidity and poor asthma control. Neuronal and inflammatory pathways have been shown to link upper and lower airway disease activity (Braunstahl et al. 2006). Acid reflux may affect asthma activity via direct aspiration into airways or stimulation of nerves innervating the

FIGURE 4–1a. STEPWISE APPROACH FOR MANAGING ASTHMA IN CHILDREN 0–4 YEARS OF AGE

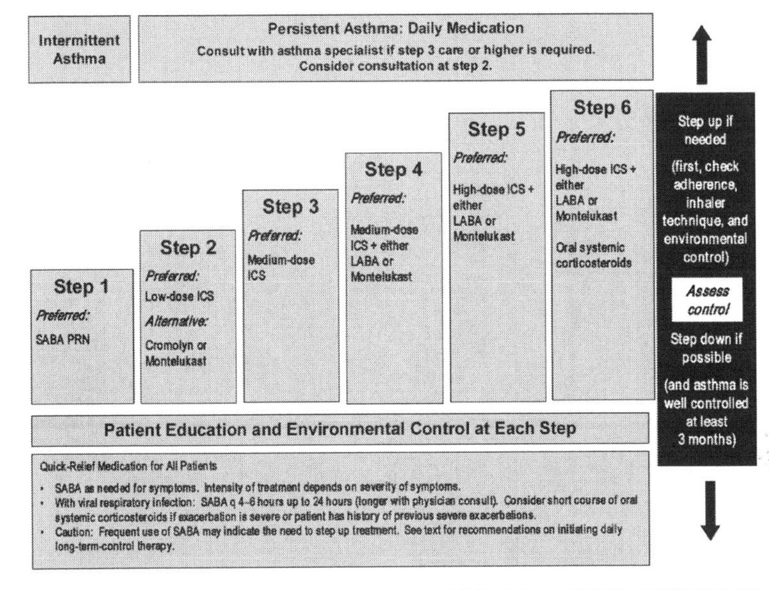

FIGURE 4–1b. STEPWISE APPROACH FOR MANAGING ASTHMA IN CHILDREN 5–11 YEARS OF AGE

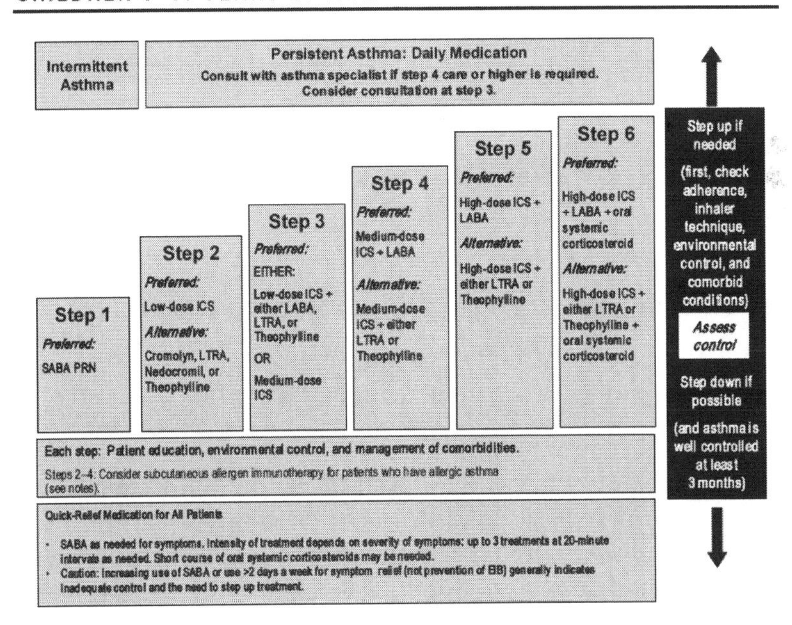

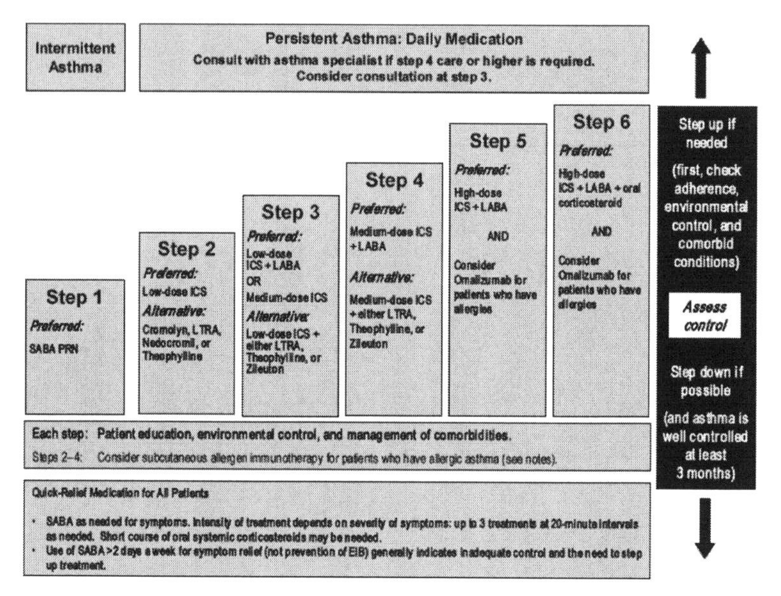

Figure 4 U.S. NHLBI Asthma Guidelines Step Up and Step Down Approach to Therapy.
NHLBI 2007 Guidelines for the diagnosis and management of asthma Expert Panel Report 3. Public domain document.
Color image of this figure appears in the color plate section at the end of the book.

lungs. Treatment of these conditions often results in improved asthma control, reduced risk of exacerbations and reduced medication requirements (Kiljander et al. 1999, Corren et al. 2004).

PHARMACOTHERAPY FOR LONG TERM ASTHMA CONTROL IN CHILDREN

Since asthma is a chronic inflammatory disease, anti-inflammatory medications or asthma controllers remain the cornerstone of long term asthma therapy. U.S. and global asthma guidelines recommend a dynamic approach to therapy dependent upon asthma control (EPR-3 2007, GINA 2002). Medication is stepped up when control is poor and stepped down when control is sustain for extended periods of time as previously described.

INHALED CORTICOSTEROIDS

Inhaled corticosteroids (ICS) are the most potent and effective anti-inflammatory medication available and mainstay of asthma therapy in children. Treatment with ICS attenuates late-phase reaction to allergens, airway hyper-responsiveness, and inflammatory cell migration/activation. ICS therapy reduces asthma symptoms, improve quality of life, and reduces of asthma-related morbidity and mortality. In contrast to systemic corticosteroids, side effects of ICS are small and often outweighed by their efficacy. Oropharyngeal candidiasis and dysphonia are the topical side effects most commonly reported. Prolonged use of high dose ICS may result in systemic adverse effects including easy bruising, adrenal suppression, and decreased bone mineral density, although this often depends on potency, bioavailability and dose of the particular ICS (Lipworth et al. 1999). Risk of side effects is small in patients receiving low doses ICS, and do not appear to affect bone mineral density or increase incidence of subcapsular cataracts in children. Small reduction in linear growth velocity occurs consistently at medium and high doses of ICS and has been reported in some children receiving low doses of ICS (CAMP 2000, Guilbert et al. 2006). Despite the loss of initial growth velocity, there is evidence that ultimate height is not affected in children receiving ICS long term (Agertoft et al. 2000). Studies however have demonstrated repeatedly that small reduction in growth velocity is far outweighed by significant improved asthma outcomes associated with treatment using low to medium dose ICS.

In the Childhood Asthma Management Program (CAMP) study, 1041 children from 5 to 12 years of age with mild-to-moderate asthma were randomly assigned to receive either budesonide ICS, nedocromil or placebo, and followed long term (mean 4.3 years). Patients receiving budesonide had asthma related morbidity compared to those who took placebo or nedocromil. Side effects of budesonide included a small, transient reduction in growth velocity. Mean increase in height in the budesonide group was 1.1 cm less than the mean increase in the placebo group. The difference between the budesonide and placebo groups in the rate of growth was evident primarily within the first year of treatment.

Budesonide ICS safety and tolerability was assessed in the START study in the long-term treatment of children and adults with persistent asthma (Sheffer et al. 2005). During the first 3 years of study, patients receiving budesonide had a 44% reduction in risk of first severe asthma attack compared to patients on placebo. Similar to CAMP, children receiving ICS experienced small decreases in growth velocity versus those receiving placebo.

The long term safety of ICS in children was studied in one of the longest studies with a mean duration of 9.2 years (Agartoff et al. 2000). No adverse effect on final adult height was found in 211 children receiving medium doses budesonide over this period of time. During the first years of treatment, there were decreases in growth velocity but these changes in early growth rate were not significantly associated with reductions in predicted final adult height.

INHALED CORTICOSTEROIDS AND AIRWAY REMODELING

Asthma patients experience airway remodeling including basement membrane thickening, mucous hyper-secretion and airway hyper-responsiveness. Previous guidelines speculated that ICS treatment early could delay or attenuate airway remodeling in children with asthma. Increasing evidence however has shown that ICS do not appear to change the natural progression of asthma including airway remodeling.

Using post bronchodilator forced expiratory volume in 1 second (FEV_1) to assess effectiveness of budesonide on airway remodeling the CAMP study demonstrated that initial post bronchodilator FEV1 was higher in children receiving the ICS compared to those taking nedocromil and placebo but there was no difference between the three groups at the end of the study (CAMP 2000). It was possible that treatment was started too late after the onset of asthma and thus airway remodeling had already occurred. This issue was addressed in the Prevention in Early Asthma in Kids wherein ICS therapy was initiated in patients 2 to 3 years of age with early onset wheezing but at high risk of persistent asthma based upon a modified asthma predictive index (Guilbert et al. 2006). During the treatment period of 2 years, patients who received ICS had a higher proportion of symptom free days, lower asthma exacerbation rates and decreased use of supplementary controller medication compared to patients given placebo. However in the 1 year observation period that followed when all patients were placed on placebo, there was no difference between the two groups. There were no differences in lung function between ICS and placebo treated patients at the end of the study.

COMBINATION THERAPY

Patients who do not achieve asthma control with low dose ICS require step up therapy to medium dose ICS or combination ICS plus add-on therapy. There is evidence that the same degree of asthma control can be achieved with a lower dose of ICS plus a second agent compared to a higher dose of ICS alone (Pauwels et al. 1997, Tal et al. 2002). Combination therapy thus allows asthma control using lower doses of ICS.

INHALED CORTICOSTEROIDS IN COMBINATION WITH LONG ACTING B_2-AGONISTS

In adult asthmatics, the addition of LABAs to ICS has been shown to be more effective in controlling asthma symptoms, improving lung function and reducing risk of exacerbations versus doubling the dose of ICS (Pauwels et al. 1997). This is congruent with results from smaller pediatric studies.

Tal et al. randomized asthmatic children (mean age 11 years) with moderate persistent asthma to receive budesonide/formoterol 90/9 BID or budesonide 200 mcg BID over 12 weeks. Patients who received combination therapy had nearly doubled morning and evening peak expiratory flow rates (PEFR) compared to similar patients receiving mono ICS therapy. In a similar study investigators compared the efficacy and safety of combination therapy with budesonide and formoterol versus budesonide alone in 630 children ages 4 to 11 years in a 12 week double-blind study (Pohunek et al. 2006). The budesonide/formoterol combination significantly improved pulmonary functions compared with budesonide alone.

SAFETY OF LONG ACTING B_2-AGONISTS (LABAs)

Some adult studies have questioned the safety of LABAs. In the Salmeterol Multicenter Asthma Research Trial, daily treatment with salmeterol or placebo added to standard asthma therapy for adults resulted in an increased risk of asthma-related deaths in patients treated with salmeterol (Nelson et al. 2006). This and other similar studies led the U.S. Food and Drug Administration (FDA) to place a "black box" warning on preparations containing LABAs. A genetic polymorphism involving Arg-Arg versus Gly-Gly substitution in the β2 agonist receptor gene has been associated with worsening asthma in a subset of patients receiving salmeterol (Israel et al. 2004). However a large retrospective study involving patients receiving fluticasone/salmeterol combination showed no difference in asthma control between patients with either genotype (Bleecker et al. 2007). Similar studies have not been replicated with children. U.S. and global guidelines still recommend the use of ICS/LABA combinations for treatment of childhood asthma. Close monitoring for loss of asthma control is needed when treating children with ICS/LABA combinations. LABAs should never be used alone as mono-therapy.

INHALED CORTICOSTEROIDS IN COMBINATION WITH LEUKOTRIENE RECEPTOR ANTAGONISTS (LTRAS)

LTRAs may be used as an alternative, not preferred, treatment option for mild persistent asthma. LTRAs can also be used as adjunctive therapy with ICS, but are not preferred addition therapy for children > 12 years of age. Most studies on the use of ICS/LTRA combinations have been done with adult patients, with few studies on pediatric patients. In a multi-centered pediatric study, the addition of LTRA improved asthma control in 182 asthmatic children ages 6–11 years who previously were not adequately controlled on ICS alone (Lemanske et al. 2010).

OTHER CONTROLLER MEDICATIONS

Inhaled sodium cromoglycate, nedocromil and theophylline are also long term anti-inflammatory controller medications used for the treatment of asthma. These medications may be used as alternative mono-therapy in mild persistent asthma and in combination with ICS in more severe asthma, although this is not the preferred treatment. Cromolyn and nedocromil have excellent safety profiles, but they require dosing multiple times a day and have minimal benefit (van der Wouden et al. 2008). Theophylline is less desirable because of its narrow therapeutic window, drug interactions, and side effect profile.

TREATMENT OF ACUTE ASTHMA SYMPTOMS AND EXACERBATIONS

Triggers such as allergens and irritants result in acute asthma exacerbations which may be severe in patients with poorly controlled asthma not on long term controller medication. Exposure to environmental pollutants and proximity to freeways have been associated with poor asthma control (Huynh et al. 2010). Acute symptoms include wheezing, chest tightness, dyspnea and cough. Severe and untreated asthma exacerbations may result in initial hypoxia, respiratory fatigue, later hypercapnia, and ultimately respiratory failure. Bronchodilator medications such as albuterol, lev-albuterol and pirbuterol remain the mainstay in the treatment of acute asthma symptoms. Selective short acting β2-agonists (SABAs) are preferable as they have effective bronchodilation with less side effects.

Repetitive or continuous use of SABAs is the most effective means of reversing airflow obstruction. In general, 3 administrations of SABAs at 20–30 minute intervals are well tolerated. Frequency beyond this depends on degree of bronchodilation achieved and severity of side effects. Onset of action usually is less than 5 minutes. When high dose SABAs are required, signs of cardiac toxicity should be monitored. In mild to moderate exacerbations, SABAs delivered by metered dose inhaler with large volume space holding chamber or spacer has been shown to be similar in efficacy to nebulized SABA delivery. Nebulized SABA delivery however may be preferable in patients unable to coordinate MDI technique due to respiratory distress, age, poor coordination, etc. Majority of patients achieve sufficient reversal of airflow obstruction to be discharged from the ED after 3 doses of SABAs. Addition of ipratropium bromide to

SABAs results in additional bronchodilation, and reduces risk of hospitalization (Rodrigo et al. 2005). Systemic corticosteroids are often required in patients experiencing asthma exacerbations. Oral corticosteroids such as prednisone have been shown to be similar in efficacy as intravenous corticosteroids. Oral agents have the advantage of low cost, ease of administration and are less invasive. Patients should be given 5–10 days of systemic corticosteroids after an exacerbation to prevent relapse. Supplemental oxygen should be administered to keep oxygen saturations above 90%. Detailed discussion regarding acute management of asthma is beyond the scope of this chapter but asthma guidelines provide up-to-date reference (EPR-3 2007, GINA 2002).

ASTHMA SELF MANAGEMENT PLAN

Long term treatment of childhood asthma should include asthma self management plans. Specific details differ between different guidelines but general principles apply to most. These usually include recognition of asthma triggers specific to the patient, recognition of deterioration of asthma symptoms, emergency treatment plan and when to seek medical attention. The most popular plans include 3 zone systems of green, yellow and red corresponding to good, worsening and poor asthma control (Fig. 5). Yellow zone interventions usually require initiation of SABAs while patients who fall into the red zone usually require SABA treatment and urgent medical attention.

EXERCISE INDUCED ASTHMA

Found in 40 to 90% of asthmatics, exercise induced asthma (EIA) is airway narrowing due to a non-pharmacologic and non-immune triggers brought on by physical exertion. Current studies implicate cooling and rapid rewarding of inspired air as the mechanism for broncho-constriction (Anderson et al. 2008). Airways usually dilate during exercise but airway obstruction ensues when the workload

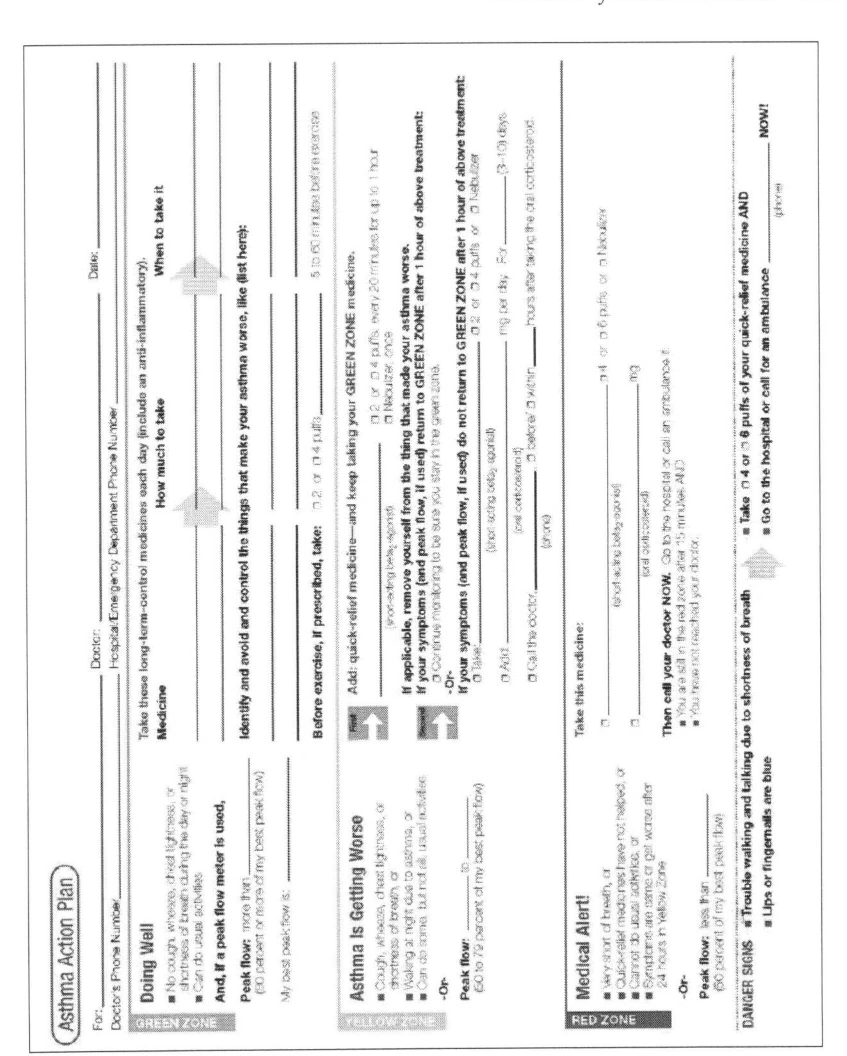

Asthma Action Plan

For: ______ Doctor: ______ Date: ______
Doctor's Phone Number ______ Hospital/Emergency Department Phone Number ______

GREEN ZONE

Doing Well

- No cough, wheeze, chest tightness, or shortness of breath during the day or night
- Can do usual activities

And, if a peak flow meter is used,

Peak flow: more than ______ (80 percent or more of my best peak flow)

My best peak flow is: ______

Take these long-term-control medicines each day (include an anti-inflammatory).

Medicine | **How much to take** | **When to take it**

Identify and avoid and control the things that make your asthma worse, like (list here):

Before exercise, if prescribed, take: ☐ 2 or ☐ 4 puffs ______ 5 to 60 minutes before exercise

YELLOW ZONE

Asthma Is Getting Worse

- Cough, wheeze, chest tightness, or shortness of breath, or
- Waking at night due to asthma, or
- Can do some, but not all, usual activities

-Or-

Peak flow: ______ to ______ (50 to 79 percent of my best peak flow)

First: **Add: quick-relief medicine—and keep taking your GREEN ZONE medicine.**
______ (short-acting beta₂-agonist) ☐ 2 or ☐ 4 puffs, every 20 minutes for up to 1 hour ☐ Nebulizer, once

Second: **If applicable, remove yourself from the thing that made your asthma worse.**
If your symptoms (and peak flow, if used) return to GREEN ZONE after 1 hour of above treatment:
☐ Continue monitoring to be sure you stay in the green zone.
-Or-
If your symptoms (and peak flow, if used) do not return to GREEN ZONE after 1 hour of above treatment:
☐ Take: ______ (short-acting beta₂-agonist) ☐ 2 or ☐ 4 puffs or ☐ Nebulizer
☐ Add: ______ (oral corticosteroid) mg per day For ______ (3–10) days
☐ Call the doctor ______ (phone) ☐ before/ ☐ within ______ hours after taking the oral corticosteroid.

RED ZONE

Medical Alert!

- Very short of breath, or
- Quick-relief medicines have not helped, or
- Cannot do usual activities, or
- Symptoms are same or get worse after 24 hours in Yellow Zone

-Or-

Peak flow: less than ______ (50 percent of my best peak flow)

Take this medicine:
☐ ______ (short-acting beta₂-agonist) ☐ 4 or ☐ 6 puffs or ☐ Nebulizer
☐ ______ (oral corticosteroid) mg

Then call your doctor NOW. Go to the hospital or call an ambulance if:
- You are still in the red zone after 15 minutes AND
- You have not reached your doctor.

DANGER SIGNS
- **Trouble walking and talking due to shortness of breath**
- **Lips or fingernails are blue**

- **Take ☐ 4 or ☐ 6 puffs of your quick-relief medicine AND**
- **Go to the hospital or call for an ambulance** ______ (phone) **NOW!**

Figure 5 Sample Asthma Self Management Plan.
NHLBI 2007 Guidelines for the diagnosis and management of asthma Expert Panel Report 3. Public domain document.

is completed and peaks in 5 to 10 minutes. There is spontaneous resolution in usually 30 minutes. EIA symptoms are not used to classify asthma severity or control. Warm up before exercise helps attenuate EIA episodes. Popular agents effective and well tolerated for prophylaxis of EIA include SABAs, LTRAs and cromoglycate. ICS are not effective given as single dose before exercise but chronic administration reduces EIA symptoms (Atkinson et al. 2003).

ALLERGIC AND NON ALLERGIC TRIGGERS OF ASTHMA

Viral infections remain main cause of asthma exacerbations in children. Environmental irritants such as fumes also trigger asthma attacks especially in patients with airway hyper-reactivity. Majority of asthmatic children are atopic to perennial allergens such as dust mites, mold, pets and cockroach and seasonal allergens including trees, grasses and weeds. Exposure to these may result in immunologic trigger of asthma symptoms via IgE mediated mast cell activation. Avoidance of these allergic triggers with strategies such as dust mite proof covers and cleaning may reduce risk of exacerbations. Allergen immunotherapy may help in controlling asthma symptoms. Detailed discussion is beyond the scope of this review and the reader is referred to several comprehensive resources (Cox et al. 2011).

Key Facts, Asthma Statistics

- Prevalence of asthma has increased worldwide over the last few decades.
- In the U.S. the prevalence increased 75% between 1980 and 1994.
- WHO estimates that 300 million people suffer from asthma worldwide.
- WHO estimates that by 2025 an additional 100 million people will be afflicted by asthma.
- WHO estimates that in 2005 255,000 people died from asthma
- 13 million school days are lost in the U.S. due to asthma.
- Asthma is the third leading cause of U.S. hospitalizations of children under 15 years.
- Asthma accounts for 217,000 emergency room visits and 10.5 million physician office visits every year in the U.S.

Key Facts, Asthma Symptoms

- Asthma is manifested by different symptoms.
- Principle pathology responsible for asthma symptoms are airway obstruction, chronic airway inflammation, mucous production and airway hyperresponsiveness.
- Airway obstruction may result in audible wheezing and chest tightness.
- Mucous production and irritation of the airways may result in a cough that may be productive.
- Worsening asthma results in inability of the lungs to oxygenate and eliminate carbon dioxide resulting in dyspnea and respiratory distress.
- Increased airway hyperreactivity is often found in people with uncontrolled asthma resulting in their airways to react and narrow in response to non-specific triggers such as tobacco smoke, and air pollutants.
- Acute asthma symptoms are relieved with bronchodilators (relievers).
- Long term prevention of asthma symptoms is controlled by anti-inflammatory medications (controllers).

Key Facts, Challenges of Treating Asthma

- Despite new advances in asthma medications many children suffer from asthma symptoms and attacks.
- Asthma is often difficult to diagnose because there is no simple test to diagnose the disease such as blood glucose in diabetes
- Treatment can be complex requiring at least 2 to 3 different medications given at different frequencies.
- Asthma medications are usually delivered by a metered dose inhaler, dry powder inhaler or nebulizer. Techniques for these devices may be difficult for young children or children who are not coordinated.

- Asthma is a chronic disease and can wax and wane.
- Compliance to treatment is difficult in chronic diseases as patients often stop their controller medicines when their diseases are in control.
- Non-compliance to long term asthma controller medications is the number one cause of loss of asthma control.
- There are multiple triggers of asthma symptoms and many patients may not be aware of their own specific sets of triggers.

Practice and Procedures

Spirometry Procedure

1. Explain to the patient how the forced expiratory maneuver is performed:
   a. Take the deepest possible breath after breathing in and out normally for several seconds.
   b. Bring the spirometer tube to the mouth and place it on top of the tongue between the teeth. Put the mouth firmly around the mouthpiece, making sure not to purse the lips as if blowing a musical instrument. Instruct the patient not to inhale from the mouthpiece unless information on inspiration as well as expiration is to be recorded.
   c. Keep the chin slightly elevated and make sure that the tongue is out of the mouthpiece.
   d. Without further hesitation, BLOW into the mouthpiece of the spirometer as hard, fast and completely as possible.
   e. Keep blowing as long as you can or until you are told to stop.
2. Always demonstrate for the patient the proper technique using a mouthpiece.
3. Check to see if the patient has any questions.
4. Coach the Patient.
   a. Instruct the patient: "Whenever you are ready, take the deepest possible breath, place your mouth firmly around the mouthpiece, and without further hesitation, blow into the

spirometer as hard, fast, and completely as possible." Watch the subject inhale fully, place the mouthpiece, and BLOW out the air.

b. Actively and forcefully coach the patient as he/she performs the maneuver. Emphasize, "BLOW" the air out, blow, keep blowing, keep blowing!"
c. Keep coaching them to continue to exhale until the point at which the tracing becomes almost flat—an obvious plateau in the volume-time curve. Since the end of test is hard for the technician to determine during the maneuver, tell the patient to blow as long as he/she can. After each maneuver let him/her relax for a few minutes.

SUMMARY POINTS

- Diagnosis of asthma is made using a combination of clinical and objective measures like spirometry.
- Many young children who wheezed especially in response to viral infections do not have persistent asthma during school age years.
- Children who have atopy are more likely to develop persistent asthma.
- Asthma control is the long term goal of asthma treatment.
- Achieving asthma control reduces symptoms and risk of exacerbations.
- Asthma control is measured by two domains: impairment and risk.
- Improved compliance and treatment of co-morbid conditions improves asthma control.
- Anti-inflammatory "controller" medications are mainstay in treatment of persistent asthma with inhaled corticosteroids as first line in most categories of persistent asthma.
- Short acting beta agonists are mainstay in treatment of acute asthma symptoms.
- Controller medications are stepped up when control is lost and stepped down when control is sustained for a period of time.

- Inhaled corticosteroids improve asthma symptoms and reduce risk of asthma attack but do not prevent airway remodeling in asthma patients.
- Small reductions in growth velocity occurs with initiation of inhaled corticosteroid activity but usually catch up growth occurs.
- An asthma self management plan help patients manage exacerbations.

DEFINITIONS AND EXPLANATION OF WORDS AND TERMS

Asthma: Chronic airway disease that is characterized by waxing and waning symptoms of wheezing, chest tightness, shortness of breath and cough. Chronic inflammation is believed to be central to the development of asthma and treatment is with long term use of anti-inflammatory medications.

Spirometry: Also known as lung function testing. Measures various physical parameters of the lung and its function. This includes measuring degree of airway narrowing in asthma patients. Other measures such as lung volumes can also be measured.

Forced Expiratory flow in 1 second (FEV_1): This is the volume of air that is maximally forced out of the lungs in 1 second. This spirometric measurement is a good measure of airway obstruction.

Wheezing: High pitched whistling noised produced by turbulent airflow through narrowed airways in asthma and other lung diseases.

Allergic Rhinitis: Allergy and Inflammation of the nose caused by allergic triggers resulting in symptoms of sneezing, runny nose, itchy nose and nasal congestion.

Eosinophilia: Increased presence of a type of white blood cell found in patients who have allergies and other disease states such as parasites, infections and malignancies.

Gastro-esophageal reflux: Disease wherein stomach acid refluxes back up into the esophagus.

Corticosteroids: Medication that reduces inflammation in many diseases. Mechanism of action involves entry into the nucleus of cells and turning of transcription of various inflammatory genes.

Growth Velocity: Rate of linear growth in children. This is usually measured as a change in length per unit of time, for instance 5 cm per year.

Allergens: Proteins, often part of animals and plants which cause allergic reactions such as hay fever and asthma. Only exposure to small amounts of these substances are required to elicit a reaction.

Agonists: Medications/agents that bind to a receptor on cells and tissue resulting in a physiologic action or reaction of that cell or tissue.

LIST OF ABBREVIATIONS

FEV_1	:	Forced expiratory volume in 1 second
FVC	:	Forced vital capacity
GERD	:	Gastroesophageal reflux
ICS	:	Inhaled corticosteroids
LABA	:	Long acting beta agonist
LTRA	:	Leukotriene receptor antagonist
MDI	:	Metered dose inhaler
EIA	:	Exercise induced asthma

REFERENCES

Agertoft, L. and S. Pedersen. 2000. Effect of long-term treatment with inhaled budesonide on adult height in children with asthma: N Engl J Med v. 343(15), p. 1064–1069.

Anderson, S.D. and P. Kippelen. 2008. Airway injury as a mechanism for exercise-induced bronchoconstriction in elite athletes: J Allergy Clin Immuno v. 122, p. 225–235.

Adkinson, N., J. Yunginger, W. Busse, B. Bochner, S. Holgate and E. Middleton. 2003. Middleton's Allergy: Principles and Practice. 6th ed. Mosby, St. Louis, MO.

ATS. 1995. American Thoracic Society. Standardization of spirometry, 1994 update: Am J Respir Crit Care Med v. 52(3), p. 1107–1136.

Bleecker, E.R., D.S. Postma, R.M. Lawrance, D.A. Meyers, H.J. Ambrose and M. Goldman. 2007. Effect of ADRB2 polymorphisms on response to long-acting beta2-agonist therapy: a pharmacogenetic analysis of two randomised studies: Lancet v. 370, p. 2118–2125.

Braunstahl, G.J. and P.W. Hellings. 2006. Nasobronchial interaction mechanisms in allergic airway airways disease: Curr Opin Otolaryngol Head Neck Surg v. 14(3), p. 176–182.

CAMP. The Childhood Asthma Management Program Research Group, 2000, Long-term effects of budesonide or nedocromil in children with asthma: N Eng J Med v. 343(15), p. 1054–1063.

Castro-Rodriguez, J.A., C.J. Holberg, A.L. Wright and F.D. Martinez. 2000. A clinical index to define risk of asthma in young children with recurrent wheezing: Am J Respir Crit Care Med v. 162(4 Pt 1), p. 1403–1406.

Corren, J., B.E. Manning, S.F. Thompson, S. Hennessy and B.L. Strom. 2004. Rhinitis therapy and the prevention of hospital care for asthma: a case-control study: J Allergy Clin Immunol v. 113(3), p. 415–419.

EPR-3, 2007, National Asthma Education and Prevention Program, Guidelines for the Diagnosis and Management of Asthma (EPR-3), Publication No. 08-4051, National Heart Lung and Blood Institute, National Asthma Education and Prevention Program, Bethesda, MD, USA.

Gamble, J., M. Stevenson, E. McClean and L.G. Heaney. 2009. The prevalence of nonaderence in difficult asthma: Am J Respir Crit Care Med v. 180(9), p. 817–822.

GINA. 2002. Global Initiative for Asthma management and Prevention. NHLBI/WHO Workshop Report. NIH Publication No. 02-3659. Bethesda, MD: Department of Health and Human Services; National Institutes of Health; National Heart, Lung, and Blood Institute.

Guilbert, T.W., W.J. Morgan, R.S. Zeiger, D.T. Mauger, S.J. Boehmer, S.J. Szefler, L.B. Bacharier, R.F. Jr. Lemanske, R.C. Strunk, D.B. Allen, G.R. Bloomberg, G.Heldt, M. Krawiec, G. Larsen, A.H. Liu, V.M. Chinchilli, C.A. Sorkness, L.M. Taussig, and F.D. Martinez. 2006. Long-term inhaled corticosteroids in preschool children at high risk for asthma: N Engl J Med v. 354(19), p. 1985–1997.

Howell, G. 2008. Nonadherence to medical therapy in asthma: risk factors, barriers, and strategies for improving: J Asthma v. 45(9), p. 723–729.

Huynh, P.N., L.G. Scott and K.Y. Kwong. 2010. Long-term maintenance of pediatric asthma: focus on budesonide/formoterol inhalation aerosol:Ther Clin Risk Manag v. 3(6), p. 65–75.

Huynh, P., M.T. Salam, T. Morphew, K.Y. Kwong and L. Scott. 2010. Residential proximity to freeways is associated with uncontrolled asthma in inner-city Hispanic children and adolescents: Journal of Allergy, Volume (2010), Article ID 157249, 7 pages doi:10.1155/2010/157249.

Israel, E., V.M. Chinchilli, J.G. Ford, H.A. Boushey, R. Cherniack, T.J. Craig, A. Deykin, J.K. Fagan, J.V. Fahy, J. Fish, M. Kraft, S.J. Kunselman, S.C. Lazarus, R.F. Jr. Lemanske, S.B. Liggett, R.J. Martin, N. Mitra, S.P. Peters, E. Silverman, C.A. Sorkness, S.J. Szefler, M.E. Wechsler, S.T. Weiss and J.M. Drazen. 2004.Use of regularly scheduled albuterol treatment in asthma: genotype-stratified, randomised, placebo-controlled cross-over trial: Lancet v. 364(9444), p. 1505–1512.

Cox, L., H. Nelson, R. Lockey, C. Calabria, T. Chacko, I. Finegold, M. Nelson, R. Weber, D.I. Bernstein, J. Blessing-Moore, D.A. Khan, D.M. Lang, R.A. Nicklas, J. Oppenheimer, J.M. Portnoy, C. Randolph, D.E. Schuller, S.L. Spector and S. Tilles and D. Wallace. 2011. Joint Task Force on Practice Parameters, American

Academy of Allergy, Asthma and Immunology, American College of Allergy, Asthma and Immunology, Joint Council of Allergy, Asthma and Immunology. Allergen immunotherapy: a practice parameter third update: J Allergy Clin Immunol v. 127(1 Suppl), p. S1–55.

Jones, C.A., L.T. Clement, T. Morphew, K.Y. Kwong, J. Hanley-Lopez, F. Lifson, L. Opas and J.J. Guterman. 2007. Achieving and maintaining asthma control in an urban pediatric disease management program: the Breathmobile Program: J Allergy Clin Immunol v. 119(6), p. 1445–1453.

Kiljander, T.O., E.R. Salomaa, E.K. Hietanen, E.O. Terho. 1999. Gastroesophageal reflux in asthmatics: a double-blind, placebo-controlled crossover study with omeprazole: Chest, v. 116(5), p. 1257–1264.

Kwong, K.Y., T. Morphew, L. Scott, J. Guterman, C.A. Jones. 2008. Asthma control and future asthma-related morbidity in inner-city asthmatic children: Ann Allergy Asthma Immunol., v. 101(2), p. 144–1452.

Lemanske, R.F. Jr., D.T. Mauger, C.A. Sorkness, D.J. Jackson, S.J. Boehmer, F.D. Martinez, R.C. Strunk, S.J. Szefler, R.S. Zeiger, L.B. Bacharier, R.A. Covar, T.W. Guilbert, G. Larsen, W.J. Morgan, M.H. Moss, J.D. Spahn, L.M. Taussig. Childhood Asthma Research and Education (CARE) Network of the National Heart, Lung, and Blood Institute. 2010, Step-up therapy for children with uncontrolled asthma receiving inhaled corticosteroids: N Engl J Med v. 362(11), p. 975–985.

Lipworth, B.J. 1999. Systemic adverse effects of inhaled corticosteroid therapy: A systematic review and meta-analysis: Arch Intern Med v. 159(9), p. 941–955.

Nelson, H.S., S.T. Weiss, E.R. Bleecker, S.W. Yancey, P.M. Dorinsky. SMART Study Group. 2006. The salmeterol multicenter asthma research trial: a comparison of usual pharmacotherapy for asthma or usual pharmacotherapy plus salmeterol: Chest, v. 129(1), p. 15–26.

Pauwels, R.A., C.G. Ofdahl, D.S. Postma, A.E. Tattersfield, P. O'Byrne, P.J. Barnes and A. Ullman. 1997. Effect of inhaled formoterol and budesonide on exacerbations of asthma. Formoterol and Corticosteroids Establishing Therapy (FACET) International Study Group: N Engl J Med v. 337(20), p. 1405–1411.

Pohunek, P., P. Kuna, C. Jorup and K.D. Boeck. 2006. Budesonide/formoterol improves lung function compared with budesonide alone in children with asthma: Pediatr Allergy Immunol v. 17(6), p. 458–465.

Rodrigo, G.J. and J.A. Castro-Rodriguez. 2005. Anticholinergics in the treatment of children and adults with acute asthma: a systematic review with meta-analysis: Thorax, v. 60(9), p. 740–746.

Sheffer, A.L., M. Silverman, A.J. Woolcock, P.V. Díaz, B. Lindberg and B. Lindmark. 2005. Long-term safety of once-daily budesonide in patients with early-onset mild persistent asthma: Results of the Inhaled Steroid Treatment as Regular Therapy in Early Asthma (START) study: Ann Allergy Asthma Immunol v. 94(1), p. 48–54.

Tal, A., G. Simon, J.H. Vermeulen, V. Petru, N. Cobos, M.L. Everard and K. de Boeck. 2002. Budesonide/formoterol in a single inhaler versus inhaled corticosteroids alone in the treatment of asthma: Pediatr Pulmonol v. 34(5), p. 342–350.

van der Wouden, J.C., J.H. Uijen, R.M. Bernsen, M.J. Tasche and J.C. de Jongste, F. Ducharme. 2008. Inhaled sodium cromoglycate for asthma in children: Cochrane Database Syst Rev v. 8(4):CD002173.

Index

S

T

U

W

About the Editors

Colin R. Martin BSc PhD RN YCAP CPsychol CSci AFBPsS is a qualified Nurse and Chair in Mental Health at the University of the West of Scotland and Adjunct Professor at the Royal Melbourne Institute of Technology (RMIT), Melbourne, Australia. He is also a Chartered Health Psychologist and a Chartered Scientist and has worked in senior management posts in the NHS followed by academic posts in the UK and the Far East. He has conducted original research in both the addictions and the mental health aspects of chronic diseases. Professor Martin is honorary Consultant Psychologist to The Salvation Army, UK and Eire Territories and was instrumental in formulating the addictions policy of the Salvation Army (UK and Eire) over recent years to develop high quality and evidence-based clinical care and services. He has published many scientific papers in psychology, biology, medical and nursing journals. He is also Editor of several books.

Victor R. Preedy PhD, DSc, FSB, FRCPath, FRSPH is Professor of Nutritional Biochemistry, King's College London, Professor of Clinical Biochemistry, King's College Hospital (Honorary) and Director of the Genomics Centre, King's College London. Presently he is a member of the King's College London School of Medicine. In his career Professor Preedy has carried out research at the National Heart Hospital (part of Imperial College London) and the MRC Centre at Northwick Park Hospital. He has collaborated with research groups in Finland, Japan, Australia, USA and Germany. He is a leading expert on the mechanisms of disease and has lectured nationally and internationally. He has published in many peer-reviewed journals and has edited over 20 books.

Color Plate Section

Chapter 11

FIGURE 4–1a. STEPWISE APPROACH FOR MANAGING ASTHMA IN CHILDREN 0–4 YEARS OF AGE

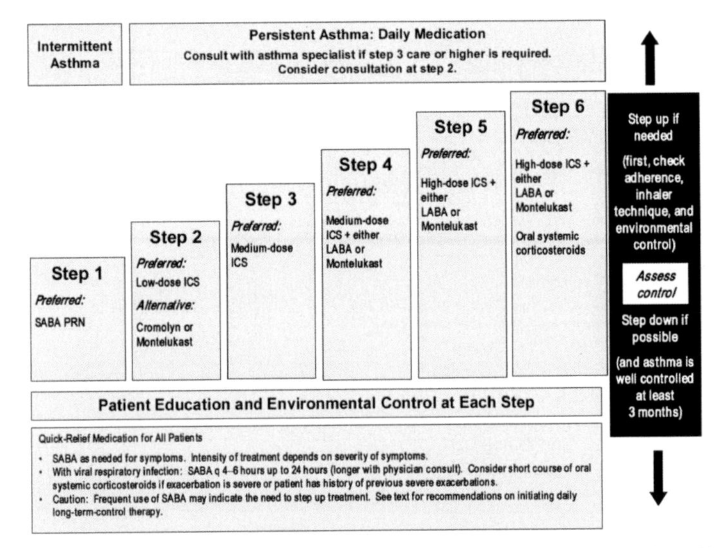

FIGURE 4–1b. STEPWISE APPROACH FOR MANAGING ASTHMA IN CHILDREN 5–11 YEARS OF AGE

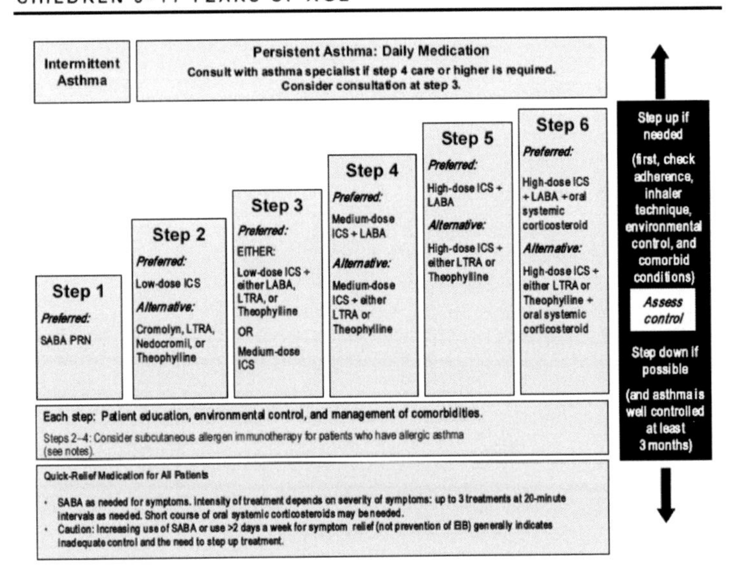

FIGURE 4–5. STEPWISE APPROACH FOR MANAGING ASTHMA IN YOUTHS ≥12 YEARS OF AGE AND ADULTS

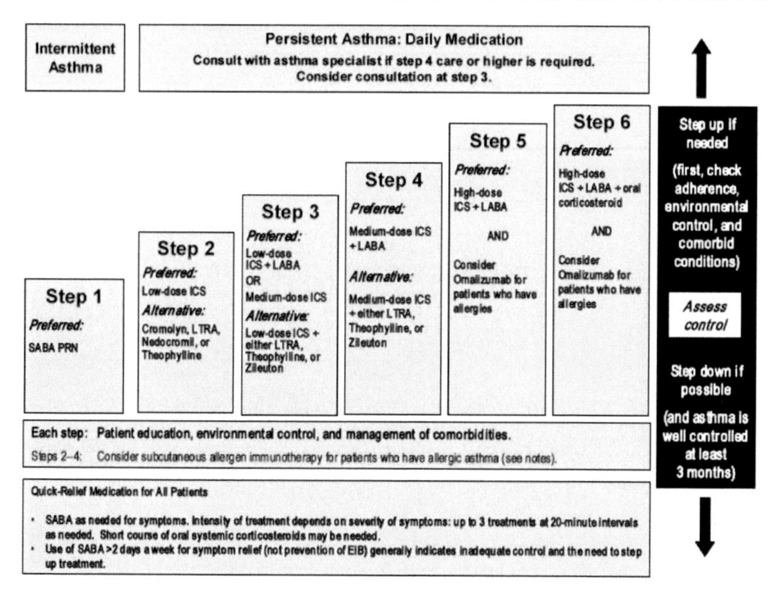

Figure 4 U.S. NHLBI Asthma Guidelines Step Up and Step Down Approach to Therapy.
NHLBI 2007 Guidelines for the diagnosis and management of asthma Expert Panel Report 3. Public domain document.